Communications
in Computer and Information Science   459

T0212586

Valeri Mladenov   Chrisina Jayne
Lazaros Iliadis (Eds.)

Engineering Applications of Neural Networks

15th International Conference, EANN 2014
Sofia, Bulgaria, September 5-7, 2014
Proceedings

Springer

Volume Editors

Valeri Mladenov
Technical University of Sofia, Bulgaria
E-mail: valerim@tu-sofia.bg

Chrisina Jayne
Coventry University, UK
E-mail: ab1527@coventry.ac.uk.

Lazaros Iliadis
University of Thrace, Orestiada, Greece
E-mail: liliadis@fmenr.duth.gr

ISSN 1865-0929   e-ISSN 1865-0937
ISBN 978-3-319-11070-7   e-ISBN 978-3-319-11071-4
DOI 10.1007/978-3-319-11071-4
Springer Cham Heidelberg New York Dordrecht London

Library of Congress Control Number: 2014947387

Typesetting: Camera-ready by author, data conversion by Scientific Publishing Services, Chennai, India

Printed on acid-free paper

Springer is part of Springer Science+Business Media (www.springer.com)

Preface

The EANN conference promotes neural networks and associated techniques and the significant benefits that can be derived from their use. The conference is not only for reporting advances, but also for showing how neural networks provide practical solutions in a wide range of applications.

The 15th EANN conference was held at the Technical University in Sofia, Bulgaria, during September, 2014. The sponsors for the conference were the Technical University in Sofia, the International Neural Network Society (INNS), and the John Atanasoff Union of Automation and Informatics. EANN 2014 attracted delegates from 16 countries across the world: France, Greece, Cyprus, USA, Italy, UK, Brazil, Bangladesh, Turkey, Japan, China, Portugal, Bulgaria, Norway, Mexico, and Finland.

This volume includes the papers that were accepted for presentation at the conference. The papers demonstrate a variety of applications of neural networks and other computational intelligence approaches to challenging problems relevant to society and the economy. These include areas such as: environmental engineering, facial expression recognition, classification with parallelization algorithms, control of autonomous unmanned aerial vehicles, intelligent transport, flood forecasting, classification of medical images, renewable energy systems, intrusion detection, fault classification, and general engineering. All papers were subject to a rigorous peer-review process by at least two independent academic referees. EANN accepted approximately 50% of the submitted papers for full length presentation at the conference. The best ten papers will be invited to submit extended contributions for inclusion in a special issue of *Neural Computing and Applications* (Springer).

The following keynote speakers were invited and gave lectures on exciting neural network application topics:

- Professor Nikola Kasabov, Director and Founder, Knowledge Engineering and Discovery Research Institute (KEDRI), Chair of Knowledge Engineering, Auckland University of Technology New Zealand
- Professor Marley Vellasco, Head of the Electrical Engineering Department and the Applied Computational Intelligence Laboratory (ICA) at PUC-Rio, Brazil
- Professor John MacIntyre, Dean of the Faculty of Applied Sciences, Pro Vice Chancellor Director of Research, Innovation & Employer Engagement, University of Sunderland, UK
- Professor Jun Wang, IEEE Fellow, Faculty of Engineering, The Chinese University of Hong Kong
- Professor Xin Yao, IEEE Fellow, School of Computer Science at the University of Birmingham Director of the Centre of Excellence for Research in Computational Intelligence and Applications (CERCIA), UK

On behalf of the conference Organizing Committee, we would like to thank all those who contributed to the organization of this year's program, and in particular the Program Committee members.

September 2014   Valeri Mladenov
                                                                  Chrisina Jayne
                                                                  Lazaros Iliadis

Organization

General Chairs

Valeri Mladenov   Technical University Sofia, Bulgaria
Chrisina Jayne   Coventry University, UK

Honorary Chair

Nikola Kasabov   Auckland University of Technology, New Zealand

Program Chairs

Valeri Mladenov   Technical University Sofia, Bulgaria
Chrisina Jayne   Coventry University, UK
Lazaros Iliadis   Democritus University of Thrace, Greece

Local Organizing Committee

Georgi Tsenov   Technical University Sofia, Bulgaria
Agata Manolova   Technical University Sofia, Bulgaria
Yancho Todorov   Technical University Sofia, Bulgaria
Stanislav Panev   Technical University Sofia, Bulgaria

Program Committee

A. Gegov   University of Portsmouth, UK
A. Andreou   University of Cyprus, Cyprus
A. Adamopoulos   Democritus University of Thrace, Greece
A. Tsitiridis   University of Swansea, UK
A. Likas   University of Ioannina, Greece
A. Tsadiras   Aristotle University of Thessaloniki, Greece
A. Tefas   Aristotle University of Thessaloniki, Greece
B. Reljin   University of Belgrade, Serbia
C. Schizas   University of Cyprus
C. Christodoulou   University of Cyprus
C. Moschopoulos   KU Leuven, Belgium
D. Nkantah   Coventry University, UK

D. Iakovidis	Technological Educational Institute of Central Greece
E. F. Georgopoulos	Technological Educational Institute of Kalamata, Greece
E. Kyriacou	Frederick University, Cyprus
E. Pimenidis	University of East London, UK
E. Papatheocharous	University of Cyprus
F. Doctor	Coventry University, UK
F. Carlo Morabito	Università degli Studi Mediterranea di Reggio Calabria, Italy
F. Marcelloni	University of Pisa, Italy
G. Magoulas	University of London, UK
G. Gnecco	University of Genova, Italy
G. Sermpinis	Glasgow University, UK
G. Beligiannis	University of Patras, Greece
H. Papadopoulos	Frederick University, Cyprus
H.-J. von Mettenheim	University Hannover, Germany
H. Pérez-Sánchez	University of Murcia, Spain
I. Maglogiannis	University of Piraeus, Greece
I. Karydis	Ionian University, Greece
I. Valavanis	Technical University of Athens, Greece
I. Bukovsky	Czech Technical University in Prague, Czech Republic
J. F. De Canete Rodriguez	University of Malaga, Spain
K. Karatzas	Aristotle University of Thessaloniki, Greece
K. A. Theofilatos	University of Patras, Greece
M. Sanguineti	University of Genoa, Italy
M. Malcangi	University of Milan, Italy
M. Elshaw	Coventry University, UK
M. Fiasche	Politecnico di Milano, Italy
M. Odetayo	Coventry University, UK
N. Shah	Coventry University, UK
N. Mitianoudis	Democritus University of Thrace, Greece
P. Božek	Slovak University of Technology, Slovak Republic
P. Kumpulainen	Tempere University of Technology, Finland
P. Koprinkova-Hristova	Bulgarian Academy of Sciences, Bulgaria
P. Hajek	University of Pardubice, Czech Republic
P. Angelov	Lancaster University, UK
R. Iqbal	Coventry University, UK
R. J. Naharro	Universidad de Huelva, Spain

R. Kamimura   Hiratsuka Kanagawa, Japan
R. Rosillo   Rafael Rosillo, University of Oviedo, Spain
S. Yue   University of Lincoln, UK
S. W. Lee   University of East London, UK
S. Likothanasis   University of Patras, Greece
V. Verykios   University of Thessaly, Greece
V. Kurkova   Czech Academy of Sciences
V. Zamudio-Rodíguez   Spain

Sponsoring Organizations

Technical University, Sofia
International Neural Network Society (INNS)
John Atanasoff Union of Automation and Informatics

Table of Contents

Fuzzy Inference ANN Ensembles for Air Pollutants Modeling in a Major Urban Area: The Case of Athens

Ilias Bougoudis, Lazaros Iliadis, and Antonis Papaleonidas

Democritus University of Thrace,
Department of Forestry & Management of the Environment & Natural Resources,
193 Pandazidou st., 68200 N Orestiada, Greece
ibougoudis@yahoo.gr, liliadis@fmenr.duth.gr, papaleon@sch.gr

Abstract. All over the globe, major urban centers face a significant air pollution problem, which is becoming worse every year. This research effort aims to contribute towards real time monitoring of air quality, which is a target of great importance for people's health. However, a serious obstacle is the high percentage of erroneous or missing data which is highly prolonged in many of the cases. To overcome this problem and due to the individuality of each residential area of Athens, separate local ANN had to be developed, capable of performing reliable interpolation of missing data vectors on an hourly basis. Also due to the need for hourly overall estimations of pollutants in the wider area of a major city, ANN ensembles were additionally developed by employing four existing methods and an innovative fuzzy inference approach.

Keywords: ANN ensembles, Fuzzy Inference System, Air Pollution.

1 Introduction

1.1 Aim of This Research

In modern cities, industrialization and increase in the number of vehicles, combined with heating systems of residences have lead to serious problems related to air quality. Increased concentrations in the levels of near surface ozone O_3, CO, CO_2, NO, NO_2, SO_2 are currently matters of great concern. Particulate matters, also known as particle pollution or PM, are complex mixtures of extremely small particles and liquid droplets. They comprise of a number of components, including acids (such as nitrates and sulfates), organic chemicals, metals, and soil or dust particles. The most dangerous particles are the ones that have a diameter of 10 micrometers PM_{10} or smaller, because they are inhaled, they pass through the throat and nose and enter the lungs. They can cause serious health problems in a long term scale. There are also particles with a diameter equal to 2.5 micrometers which are found in smoke or haze. They are emitted from forest fires, fire places, or they can emerge from reactions of gases coming directly from industries and automobiles [31].

This research effort is twofold. First, it describes the development of reliable local ANN models, related to various measuring stations in Athens, capable of estimating the concentrations of the following air pollutants: O_3, CO, NO, NO_2, SO_2, PM_{10}, $PM_{2.5}$. Second, it presents the implementation and application of ANN ensembles for

V. Mladenov et al. (Eds.): EANN 2014, CCIS 459, pp. 1–14, 2014.

the estimation of the concentrations of the above air pollutants, in the wider area of Athens. An ensemble consists of a set of individually trained networks, whose predictions are combined when fed with novel instances. Previous research has shown that an ensemble is often more accurate than any of the single local models [10]. This is achieved by employing not only four existing ANN ensemble approaches but by introducing an innovative Fuzzy Inference Ensemble (FIE) model. It is the first time that FIE is used towards ensembles development. In this way a range of overall pollution indices are established (each one for each pollutant) that correspond to the greater area of a major urban center.

1.2   Literature Review

Several original research papers are published in the literature, dealing with Soft Computing Techniques, applied for air pollutants' estimation. The common disadvantage of these approaches is the fact that they are developing overall neural network models based mainly on data from the narrow city center, and they suppose that they can serve as the optimal ones for wide range areas belonging to wide urban or semi-urban areas [22], [23], [16], [3]. However, a more rational consideration would require the division of these cities to smaller more homogenous frames. For example the "*Athinas*" measuring station in Athens is located in the actual city center and it belongs to a very different area from "*Penteli*", in terms of vehicles' traffic, topographic plus morphological details and as far as the density of population is concerned. Thus, only local ANN models can serve as reasonable estimators in different regions of major cities.

Another limitation of many existing approaches is the fact that they have a seasonal nature by proposing models for specific times of the year (e.g. the summer months) and not for continuous long periods of time [26], [21], [14], [15], [5], [17].

This paper discusses the results of the second phase of our research effort to model air pollution in the greater area of Athens, by using Computational Intelligence [18], [19], [20]. The first stage, which started two years ago resulted in the construction of real time monitoring Multi Agent networks, plus short term ANN O_3 predictors, capable to provide 1-hour or 2-hours data when measuring stations malfunction. The depended parameter was only O_3 concentration.

In the research described here, 117 local ANN models (LANNMs) were developed to model air pollutants, each one corresponding to a measuring station, whereas the estimated pollutants were CO, NO, NO_2, SO_2, PM_{10}, $PM_{2.5}$. Moreover, ANN ensembles comprising of the 117 LANNMs, were developed following four established methods and a new FIS approach proposed here. The aim was to produce an overall model for the whole city something that has never been done so far in Greece. An ensemble with the same orientation has been developed in China [8].

2   Data and Area of Research

In particular, the data set collection of the current study was related to the wider area of Athens, for a period of 13-years (2000-2012). It comprised of hourly values for the concentration levels of CO, NO, NO_2, SO_2, PM_{10}, $PM_{2.5}$ ($\frac{\mu g}{m^3}$) as well as hourly data for air temperature (C^o), solar radiation (Wm^{-2}), wind speed ($\frac{m}{sec}$) and direction (rad), pressure (mbar), Illumine and relative humidity. Data were obtained by the Greek ministry of Environment [32]. Totally 1,017,733 data vectors without missing values were available whereas an average as high as 18.82% of the data are missing with the station of Piraeus having the worst percentage equal to 33.67%.

Table 1. Description of the stations employed for this research

ID	Station's name	Code	Missing values	Correct Data Vectors	Station's data
1	Ag. Paraskevi	AGP	12.32%	99,936	O_3, NO, NO_2, SO_2
2	Amarusion	MAR	21.58%	89,371	O_3, NO, NO_2, CO
3	Peristeri	PER	33.61%	75,668	O_3, NO, NO_2, CO, SO_2
4	Patision	PAT	10.45%	102,068	O_3, NO, NO_2, CO, SO_2
5	Aristotelous	ARI	16.76%	94,873	NO, NO_2
6	Geoponikis	GEO	26.84%	83,381	O_3, NO, NO_2, CO, SO_2
7	Piraeus	PIR	33.67%	75,600	O_3, NO, NO_2, CO, SO_2
8	N Smyrnh	SMY	26.06%	84,272	O_3, NO, NO_2, CO, SO_2
9	Penteli	PEN	3.66%	109806	Meteorological
10	Thiseion	THI	0.30%	113,632	Meteorological
11	Athinas	ATH	21.86%	89,058	O_3, NO, NO_2, CO, SO_2

The following image 1 clearly presents the exact positions of the eleven measuring stations. They are distributed in a way to cover the city center, near the coast and around the mountains of the Attica basin.

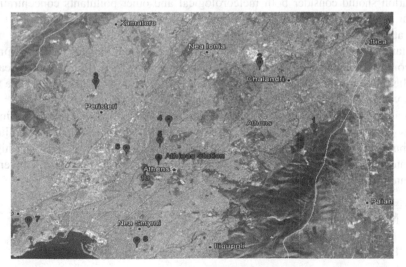

Image 1. Location of the eleven measurement stations in the wider area of Athens

3   Parameters' Selection

The choice of the involved parameters for the estimation of air pollutants was based on a combination of our own correlation analysis tests and related results obtained from the literature. The following table 2 presents a sample of the absolute correlation value of carbon monoxide and O_3 concentration in the "*Athinas*" station for the period 2005 - 2008, with available independent features.

Table 2. Sample Correlation analysis for CO concentration

Code	Parameter correlated to CO	ABS (R)	Code	Parameter correlated to O_3	ABS (R)
1	NO	0.86	8	CO	0.50
2	NO2	0.54	9	NO	0.49
3	O_3	0.49	10	NO_2	0.50
4	RH	0.24	11	RH	0.50
5	SR	0.05	12	SR	0.44
6	AIRTEMP	0.24	13	AIRTEMP	0.47
7	PR	0.33	14	WS	0.42

According to the table above, a high degree of correlation exists between carbon monoxide concentration and nitrogen dioxide. Also, it is shown that there is a fair correlation not only between O_3 in "*Athinas*" station and meteorological features, but also between O_3 with primitive pollutants of other stations as well. The above and other research efforts [21], indicate that the development of ANN for pollutants estimation should consider both meteorological and other pollutants concentrations. Similar results were obtained from the other stations. Regarding its seasonality it may seem at first sight that the need for O_3 risk monitoring and forecasting is more important during the summer period that has favorable conditions for the development of secondary pollutants. However it is self evident that this is like partially dealing with the problem that still exists during the rest of the year, especially due to the very high levels of O_3 even in months like February or March [32].

Regarding the PM_{10} concentrations, based on [24], the PM_{10} exceedances in the Athens area are related to spatial distribution characteristics and air pollution contributors. Finally according to [2], $PM_{2.5}$ and PM_{10} concentrations are highly correlated with carbon monoxide, black carbon and nitrogen oxides and inversely correlated with local wind speed.

4   Materials and Methods

4.1   Local Multilayer Feed Forward ANN

Totally 117 local ANN models using the *learngdm* learning function, the Tangent Sigmoid (*tansig*) transfer function for the hidden layer and the *purelin* for the output layer, were developed, each corresponding to one or more air pollutants' estimation in the measuring stations of Attica. The number of hidden neurons varied from 10 to 13. The training function was either *trainlm* or *trainbr*. The following tables 3, 4, 5, 6, 7 describe the performance of all 197 developed Local ANN models.

Table 3. Description and evaluation of the Local ANN models for all pollutants

All	ID	R^2	RMSE	All	ID	R^2	RMSE
1 trainlm	AGP 00-04	0.88	14.5	14 trainbr	SMY 00-04	0.63	25.0
2 trainlm	AGP 05-08	0.87	14.0	15 trainbr	SMY 05-08	0.66	23.4
3 trainlm	AGP 09-12	0.93	10.7	16 trainbr	SMY 09-12	0.76	17.8
4 trainlm	ATH 00-04	0.64	35.8	17 trainlm	PAT 00-04	0.74	35.8
5 trainlm	ATH 05-08	0.66	28.5	18 trainlm	PAT 05-08	0.74	36.1
6 trainlm	ATH 09-12	0.85	7.35	19 trainlm	PAT 09-12	0.76	28.4
7 trainlm	ARI 05-08	0.68	32.5	20 trainbr	PIR 00-04	0.64	25.7
8 trainbr	GEO 00-04	0.64	25.5	21 trainbr	PIR 05-08	0.64	21.6
9 trainbr	GEO 05-08	0.67	23.3	22 trainbr	PIR 09-12	0.76	17.8
10 trainbr	GEO 09-12	0.67	20.7	23 trainbr	PER 00-04	0.68	19.3
11 trainbr	MAR 00-04	0.63	24.4	24 trainbr	PER 05-08	0.68	18.6
12 trainbr	MAR 05-08	0.73	24.0	25 trainbr	PER 09-12	0.81	13.2
13 trainlm	MAR 09-12	0.72	19.5				

Table 4. Description and evaluation of the Local ANN models SO2

SO_2	ID	R^2	RMSE	SO_2	ID	R^2	RMSE
26 trainbr	ATH 00-04	0.64	7.03	37 trainbr	SMY 05-08	0.68	10.9
27 trainlm	ATH 05-08	0.70	4.96	38 trainbr	SMY 09-12	0.76	5.79
28 trainlm	ATH 09-12	0.75	2.79	39 trainbr	PAT 00-04	0.66	2.5
29 trainlm	ARI 00-04	0.67	9.40	40 trainlm	PAT 05-08	0.68	14.0
30 trainlm	ARI 05-08	0.74	11.1	41 trainlm	PAT 09-12	0.88	10.5

Table 4. (*Continued.*)

31 trainlm	ARI 09-12	0.68	32.5	42 trainbr	PIR 00-04	0.79	4.28
32 trainlm	GEO 00-04	0.65	7.49	43 trainbr	PIR 09-12	0.65	14.9
33 trainbr	GEO 05-08	0.64	5.01	44 trainbr	PER 00-04	0.66	2.5
34 trainbr	GEO 09-12	0.76	3.46	45 trainbr	PER 05-08	0.66	10.1
35 trainlm	MAR 05-08	0.79	11.6	46 trainbr	PER 09-12	0.73	6.84
36 trainlm	SMY 00-04	0.68	10.9				

Table 5. Description and evaluation of the Local ANN models for PM_X, NO and CO

PM_{10}	ID	R^2	RMSE	CO	ID	R^2	RMSE
47 trainbr	AGP	0.65	13.20	59 trainbr	ATH 00-04	0.90	0.60
48 trainbr	ARI	0.65	13.90	60 trainlm	GEO 00-04	0.93	0.32
49 trainbr	ARI 2	0.74	11.20	61 trainlm	GEO 05-08	0.95	0.19
50 trainbr	MAR	0.90	9.430	62 trainlm	GEO 09-12	0.93	0.16
51 trainbr	PIR	0.74	10.90	63 trainlm	SMY 00-04	0.93	0.30
PM_X	ID	R^2	RMSE	64 trainlm	SMY 05-08	0.950	0.20
52 trainlm	AGP	0.82	4.47	65 trainlm	SMY 09-12	0.930	0.18
53 trainbr	PIR	0.80	6.50	66 trainlm	PAT 00-04	0.910	0.72
$PM_{2.5}$	ID	R^2	RMSE	67 trainlm	PAT 09-12	0.920	0.28
54 trainbr	AGP	0.65	4.72	68 trainlm	PIR 00-04	0.870	0.43
55 trainbr	PIR	0.81	4.85	69 trainlm	PIR 09-12	0.930	0.18
NO	ID	R^2	RMSE	70 trainlm	PER 00-04	0.920	0.22
56 trainlm	ARI 00-04	0.680	58.3				
57 trainlm	ARI 09-12	0.760	4.75				
58 trainlm	PER 09-12	0.880	7.10				

Table 6. Description and evaluation of the Local ANN models for O_3

O_3	ID	R^2	RMSE	O_3	ID	R^2	RMSE
71 trainlm	AGP 00-04	0.81	16.70	82 trainlm	SMY 00-04	0.84	17.20
72 trainlm	AGP 05-08	0.85	12.60	83 trainlm	SMY 05-08	0.91	13.20
73 trainlm	AGP 05-08	0.86	13.50	84 trainlm	SMY 09-12	0.93	10.80
74 trainlm	ATH 00-04	0.82	13.70	85 trainlm	PAT 00-04	0.88	7.45
75 trainlm	ATH 05-08	0.91	7.95	86 trainlm	PAT 05-08	0.89	6.61
76 trainlm	GEO 00-04	0.87	12.4	87 trainlm	PAT 09-12	0.93	5.98
77 trainlm	GEO 05-08	0.91	10.8	88 trainlm	PIR 00-04	0.85	12.1
78 trainlm	GEO 09-12	0.93	9.26	89 trainlm	PIR 05-08	0.82	13.4
79 trainlm	GEO 09-12	0.93	9.43	90 trainlm	PER 00-04	0.85	15.4
80 trainlm	MAR 0004	0.83	14.7	91 trainlm	PER 05-08	0.91	11.3
81 trainlm	MAR 0508	0.88	13.5	92 trainlm	PER 09-12	0.91	10.8

Table 7. Description and evaluation of the Local ANN models for NO-NO_2

NO_X	ID	R^2	RMSE	NO_X	ID	R^2	RMSE
93 trainbr	AGP 00-04	0.76	7.73	106 trainbr	SMY 00-04	0.73	24.40
94 trainlm	AGP 09-12	0.76	4.75	107 trainlm	SMY 05-08	0.88	15.90
95 trainlm	ATH 00-04	0.90	23.50	108 trainlm	SMY 09-12	0.92	9.480
96 trainlm	ATH 05-08	0.92	18.20	109 trainlm	PAT 00-04	0.74	41.10
97 trainlm	ATH 09-12	0.88	18.60	110 trainlm	PAT 05-08	0.92	23.0
98 trainlm	ARI 00-04	0.71	39.50	111 trainlm	PAT 09-12	0.92	17.60
99 trainlm	ARI 09-12	0.93	10.70	112 trainlm	PIR 00-04	0.70	27.70
100 trainlm	GEO 00-04	0.72	30.30	113 trainlm	PIR 05-08	0.86	17.70
101 trainlm	GEO 05-08	0.92	14.40	114 trainlm	PIR 09-12	0.92	9.48

Table 7. (*Continued.*)

102 trainlm	GEO 09-12	0.77	21.50	115 trainlm	PER 00-04	0.76	17.50
103 trainbr	MAR 0004	0.91	14.40	116 trainlm	PER 05-08	0.90	11.20
104 trainlm	MAR 0508	0.92	10.70	117 trainlm	PER 09-12	0.85	8.540
105 trainlm	MAR 0912	0.93	7.01				

4.2   Proposed ANN Ensemble Model and Existing Ones

A Neural Networks' Ensemble (NNEN) comprises of a set of neural network models that reach a decision by averaging the results of the individual models by following specific algorithms. The core attribute of a NNEN is that it involves the combination of a set of ANNs, each of which accomplishes the same task (which is the case here). There is no point or advantage to combining a group of ANN that are identical and generalize in the same way. They should vary in terms of architecture, weights, convergence time, training functions and yet they should contribute to the solution of the same problem, since they produced the same pattern of errors in the testing phase [9].

A.   Fuzzy Inference Ensemble Model (FIEM)

The innovation of this paper is enhanced by the employment of a flexible approach based on fuzzy logic. For each pollutant, a Mamdani Fuzzy Inference System (FIS) has been developed. The FIS considers the range from the two evaluation metrics (correlation coefficient and mean square error). This consideration, leads to the development of corresponding Mamdani Rule Sets. From the execution of the System, two outputs are obtained, namely: *fuzzy correlation coefficient and fuzzy mean square error*. The FIEM model takes under consideration the accuracy of each network in a flexible manner. Moreover, the new values are produced through machine learning, filtered by fuzzy logic. In this way, the outputs are unbiased.

As a result, this method offers a more objective approach. When the fuzzy outputs are produced, they are averaged based on the rule set of the FIS. In most cases, the fuzzy correlation coefficient is lower than the one produced by the Simple Classic Average approach (see below), while the fuzzy mean square error is higher. A distinct Fuzzy Inference System has been developed for every pollutant. Each FIS has an Inference mechanism comprising of the following Heuristic Rule set. The differentiation between the separate systems (corresponding to each pollutant) lies in the determination of the fuzzy membership functions.

Rule set:
If (R^2 is max) and (RMSE is min) Then (R^2tot is max) AND (RMSEtot is min)
If (R^2 is min) and (RMSE is max) Then (R^2tot is min) AND (RMSEtot is max)
If (R^2 is med) and (RMSE is med) Then (R^2tot is med) AND (RMSEtot is med)
If (R^2 is min) Then (R^2tot is min)
If (R^2 is max) Then (R^2tot is max)

If (RMSE is min) Then (RMSEtot is min)
If (RMSE is max) Then (RMSEtot is max)

The first three rules were given a weight value of 0.5, whereas the last four a value of 1. This was done because in many cases the overall performance of a network is not defined by both the correlation coefficient and the root mean square error. For example, there were networks where we had a high correlation coefficient value (great for the overall performance), but also at the same time a high root mean square error (inadequate for the overall performance). So we decided that the outputs of the System should be influenced from each input separately (fuzzy correlation coefficient and fuzzy mean square error), rather than from both at the same time.

For the Fuzzy Membership Functions (FMFs), the range of each input was the range of the values for each pollutant. Finally, the FMFs used in the inputs were **Triangular** (*trimf* for MATLAB) FMF for the minimum and maximum Linguistics and **Trapezoidal** (*trapmf* in MATLAB) for the medium Linguistic whereas for the output the FMFs employed were output functions were **Triangular** for the minimum and maximum Linguistics and Gaussian (*Gausmf* in MATLAB) for the medium one. The Trapezoidal, Triangular and Gaussian FMFs can be seen in the following equations 1, 2, 3 respectively.

$$\mu_s(X) = \begin{cases} 0, & \text{if } X \le a \\ (X-a)/(m-a), & \text{if } X \in (a,m) \\ 1, & \text{if } X \in [m,n] \\ (b-X)/(b-n), & \text{if } X \in (n,b) \\ 0, & \text{if } X \ge b \end{cases} \tag{1}$$

$$\mu_s(X) = \begin{cases} 0 \text{ if } X < a \\ (X-a)/(c-a) \text{ if } X \in [a,c] \\ (b-X)/(b-c) \text{ if } X \in [c,b] \\ 0 \text{ if } X > b \end{cases} \tag{2}$$

$$f(x,\sigma,c) = e^{\frac{-(x-c)^2}{2\sigma^2}} \tag{3}$$

B. Simple Classic Average

A classical employed approach is the Simple and Classic Average (SCA) according to which the values of every pollutant for each station are simply averaged according to equation 4 [1]. $\bar{X} = \frac{1}{n}\sum_{i=1}^{n} X_i = \frac{1}{n}(X_1 + \ldots\ldots + X_n)$ (4) This approach is simple and easy to apply. However, it is not the most rational, as it does not take under consideration that some networks may be more accurate than others and thus they should be rewarded. Due to its over simplification nature, this method was used as an early approach in this research. After obtaining results from the other methods, we compared them to the ones produced by SCA.

C. Weighted Average

The third traditional method used for each pollutant is weighted average (WA). Through weights, the disadvantages of the first method are faced. All weights are given real values between 0.50 and 1. The weights are calculated as below: the best input (the maximum correlation coefficient and the minimum mean square error) is given the value 1. The second best input is given the value 0.95 and so on (following a step of 0.05). The disadvantage of this approach is that the weights are chosen arbitrarily. The averaging formula is given by equation 5 [6], [1].

$$\bar{X} = \frac{1}{n} \frac{\sum_{i=1}^{n} W_i X_i}{\sum_{i=1}^{n} W_i} \tag{5}$$

D.   Correlation Ensemble Method

The forth method uses values from the correlation matrix of each pollutant. We created such matrices under the MATLAB platform. The averaging is given by the following equations 6 and 7 [28]. $\bar{X}_{CEM} = \sum_{i=1}^{n} a_i X$ (6) where a_i is estimated by the next

equation 7. $a_i = \dfrac{\sum_{j=1}^{n} C_{ij}^{-1}}{\sum_{k=1}^{n}\sum_{j=1}^{n} C_{kj}^{-1}}$ (7) C_{ij} is the correlation between i and j input. This method

minimizes the RMSE and has produced the lowest error from all the other methods above in cases where we had the more samples. One disadvantage though, is that very small correlation values increase the output of the method.

E.   Certainty Ensemble Method

The fifth and last method uses the certainty of each network; we define certainty as the probability that a network foresees the missing value correctly. Therefore, we used the correlation coefficient of each network as its probability (equation 8).

$$C(y) = \begin{cases} y & \text{if } y \geq 0.5 \\ 1 - y, & \text{otherwise} \end{cases} \tag{8}$$

However, in our case, all the neural networks had probabilities greater than 0.63. As a result, we used the first branch of the above formula only. The averaging function and the weights' values are given by equations 9 and 10 respectively.

$$f_{DAN} = \sum_{i=1}^{n} w_i f_i(X) \tag{9}$$

$$W_i = \frac{C(f(X_i))}{\sum_{j=1}^{n} C(f_i(X))} \tag{10}$$

The transfer functions selected for the layers were sigmoid (equation 11) for the hidden layer and linear for the output layer [29]. The S-shaped logistic sigmoid function is bounded between 0 and 1, therefore input and output data should be also normalized in the same range by using equation 12.

$$f(X) = \frac{1}{1 + e^{-X}} \tag{11}$$

$$\hat{Z} = \frac{Z - Z_{min}}{Z_{max} - Z_{min}} \tag{12}$$

Data were normalized using eq. 11, where $\hat{Z}$ is the normalised value and Z_{min} and Z_{max} are the minimum and the maximum values of Z, respectively [25]. The metrics used were R^2 and RMSE (Root Mean Square Error) [4].

5   Assessment of the NNENs

The following tables 8 and 9 present the evaluation results of the developed ANN ensembles by applying the innovative FIS model (proposed here) and the existing approaches.

Table 8. Evaluation of the Ensemble ANN models for CO, NO, NO-NO_2, SO_2, O_3 in testing

CO	R^2	RMSE	NO	R^2	RMSE
SCA	0.922	0.315	SCA	0.773	23.383
FIS	0.940	0.354	FIS	0.774	25.437
Weighted Average	0.924	0.287	Weighted Average	0.785	16.787
Correlation Ensemble	0.925	0.243	Correlation Ensemble	0.773	24.900
Certainty Ensemble	0.922	0.312	Certainty Ensemble	0.782	21.337
NO – NO_2	R^2	RMSE	SO_2	R^2	RMSE
SCA	0.845	17.756	SCA	0.706	8.620
FIS	0.829	17.562	FIS	0.695	8.385
Weighted Average	0.855	16.112	Weighted Average	0.712	7.789
Correlation Ensemble	0.821	12.924	Correlation Ensemble	0.709	9.383
Certainty Ensemble	0.853	17.278	Certainty Ensemble	0.711	8.589
O_3	R^2	RMSE	O_3	R^2	RMSE
SCA	0.879	11.777	Weighted Average	0.881	11.456
FIS	0.861	11.569	Correlation Ensemble	0.909	10.832
			Certainty Ensemble	0.880	11.684

Table 9. Evaluation of the Ensemble ANN models for PM_{10}, $PM_{2.5}$, PM_X and all pollutants in testing

PM_{10}	R^2	RMSE	$PM_{2.5}$	R^2	RMSE
SCA	0.736	11.726	SCA	0.730	4.785
FIS	0.769	11.814	FIS	0.725	4.800
Weighted Average	0.752	11.482	Weighted Average	0.738	4.783
Correlation Ensemble	0.723	12.010	Correlation Ensemble	0.700	4.761
Certainty Ensemble	0.747	11.536	Certainty Ensemble	0.738	4.792
All PM	R^2	RMSE	All Pollutants	R^2	RMSE
SCA	0.810	5.485	SCA	0.722	22.538
FIS	0.800	5.675	FIS	0.710	22.867
Weighted Average	0.810	5.431	Weighted Average	0.733	21.610
Correlation Ensemble	0.812	5.229	Correlation Ensemble	0.677	22.948
Certainty Ensemble	0.810	5.472	Certainty Ensemble	0.732	21.996

From tables 8 and 9, it can be clearly seen that for CO and for PM_X the FIS has the best R^2 but the worst RMSE. For $NO-NO_2$ the FIS has the second best R^2 but the RMSE is only second worst, whereas for NO it has almost as high R^2 value as the best method (difference of the order of 10^{-3}) but the worst RMSE. For PM_{10} the FIS has the best R^2 but the not the best RMSE. Generally speaking, the differences between the five (5) used approaches for both metrics are so small that they can be considered as having equivalent performance.

6   Conclusions and Future Work

A first innovation of this research effort is the fact that it proposes a successful new FIS approach for ANN ensemble development. This approach has proven that it can perform as efficiently as the other existing methods. The second achievement of this research effort is the fact that it produces not only reliable local ANN models for various urban and semi-urban areas of a major city (Athens), but it also produces overall ANN ensembles for the same area, by employing five distinct approaches (one new and four established ones).

It has been proven that the ANN ensembles that were developed and evaluated with a huge pile of data vectors perform much better than the local ANN models in most of the cases, in terms of R^2. More specifically, in five (5) out of nine (9) cases the ensembles are more reliable, in two (2) cases we have equal values of R^2 whereas only twice the Local ANN models perform better. The following table 10 offers a strong support towards this argument.

Table 10. Comparison of Ensembles versus Local ANN in terms of R^2

Percentage of cases where Local ANN perform better for CO	Percentage of cases where Local ANN perform better for NO	Percentage of cases where Local ANN perform better for NO-NO2	Percentage of cases where Local ANN perform better for O_3	Percentage of cases where Local ANN perform better for SO_2
16.6%	33.3%	64%	60%	42.85%
Percentage of cases where Local ANN perform better for All Pollutants	Percentage of cases where Local ANN perform better for PM_{10}	Percentage of cases where Local ANN perform better for $PM_{2.5}$	Percentage of cases where Local ANN perform better for All PM	
48%	20%	50%	50%	

Future work will involve the enhancement of the proposed FIS model to use more fuzzy relations and more T-Norms and S-Norms. The aim is going to be the improvement of the system's rule set in terms of RMSE in order to overcome the contradiction where R^2 is the best achieved between all methods but RMSE remains quite high.

References

1. Baidyk, T., Kussul, E.: Ensemble Neural Networks. Optical Memory and Neural Networks 18(4), 295–303 (2009)
2. Chaloulakou, A., Kassomenos, P., Spyrellis, N., Demokritou, P., Koutrakis, P.: Measurements of PM_{10} and $PM_{2.5}$ particle concentrations in Athens. Greece Atmospheric Environment 37(2003), 649–660 (2012)
3. Hooyberghs, J., Mensink, C., Dumont, G., Fierens, F., Brasseur, O.: A neural network forecast for daily average PM10 concentrations in Belgium. Atmospheric Environment (January 2005)
4. Iliadis, L.: Intelligent Information Systems and Applications in Risk Estimation. Hrodotos Publications (2007)
5. Inal, F.: Artificial Neural Network Prediction of Tropospheric Ozone Concentrations in Istanbul, Turkey. CLEAN – Soil, Air, Water 38(10), 897–908 (2010)
6. Jimenez, D.: Dynamically weighted ensemble neural networks for classification (1998)
7. The 1998 IEEE International Joint Conference (Volume: 1)
8. Kadri, C., Tian, F., Zhang, L., Dang, L., Li, G.: Neural Network Ensembles for Online Gas Concentration Estimation Using an Electronic Nose. International Journal of Computer Science Issues 10(2(1)) (March 2013)
9. Lopez, M., Melin, P., Castillo, O.: A method for creating Ensemble Neural Networks using a Sampling Data Approach. In: Thero. Advances and Applications of Fuzzy Logic. ASC, vol. 42, pp. 772–780. Springer (2007)
10. Maclin, R., Opitz, D.: Popular Ensemble Methods: An Empirical Study. Journal of Artificial Intelligence Research 11, 169–198 (1999)
11. Mammone, R.J.: Artificial Neural Networks for Speech and Vision, pp. 126–142. Chapman & Hall, London (1993)
12. Marougianni, G.: Forecasting tropospheric ozone levels from meteorological variables: Athens urban area as a case study. Postgraduate thesis. AUTH, Greece (2010)
13. Ministry of Environment, Energy & Climate Change, Air Quality, Reports, Air Pollution 2009 Annual Report (2010)
14. Ozcan, H.K., Bilgili, E., Sahin, U., Bayat, C.: Modeling of trophospheric ozone concentrations using genetically trained multi-level cellular neural networks. Advances in Atmospheric Sciences 24(5), 907–914 (2007)
15. Ozdemir, H., Demir, G., Altay, G., Albayrak, S., Bayat, C.: Environmental Engineering Science 25(9), 1249–1254 (2008)
16. Ordieres Meré, J.B., Vergara González, E.P., Capuz, R.S., Salaza, R.E.: Neural network prediction model for fine particulate matter (PM). Environmental Modelling and Software 20, 547–559 (2005)
17. Paoli, C.: A Neural Network model forecasting for prediction of hourly ozone concentration in Corsica. In: Proceedings IEEE of the 10th International Conference on Environment and Electrical Engineering, EEEIC (2011)
18. Papaleonidas, A., Iliadis, L.: Employing ANN That Estimate Ozone in a Short-Term Scale When Monitoring Stations Malfunction. In: Jayne, C., Yue, S., Iliadis, L. (eds.) EANN 2012. CCIS, vol. 311, pp. 71–80. Springer, Heidelberg (2012a)
19. Papaleonidas, A., Iliadis, L.: Hybrid and Reinforcement Multi Agent Technology for real time air pollution monitoring. In: Iliadis, L., Maglogiannis, I., Papadopoulos, H. (eds.) Artificial Intelligence Applications and Innovations. IFIP AICT, vol. 381, pp. 274–284. Springer, Heidelberg (2012b)

20. Papaleonidas, A., Iliadis, L.: Neurocomputing techniques to dynamically forecast spatiotemporal air pollution data. Evolving Systems 4, 221–233 (2013), doi:10.1007/s12530-013-9078-5

21. Paschalidou, A., Iliadis, L., Kassomenos, P., Bezirtzoglou, C.: Neural Modeling of the Tropospheric Ozone concentrations in an Urban Site. In: Proceedings of the 10th International Conference Engineering Applications of Neural Networks, pp. 436–445 (2007)

22. Roy, S.: Prediction of Particulate Matter Concentrations Using Artificial Neural Network. Resources and Environment 2(2), 30–36 (2012), doi:10.5923/j.re.20120202.05

23. Díaz-Robles, L.A., Ortega, J.C., Fu, J.S., Reed, G.D., Chow, J.C., Watson, J.G., Moncada-Herrera, J.A.: A hybrid ARIMA and artificial neural networks model to forecast particulate matter in urban areas: The case of Temuco, Chile 42(35), 8331–8340 (2008)

24. Sfetsos, A., Vlachogiannis, D.: A new approach to discovering the causal relationship between meteorological patterns and PM_{10} exceedances. Atmospheric Research 98(2), 500–511 (2013)

25. Slini, T., Karatzas, K., Moussiopoulos, N.: Correlation of air Pollution and Meteorological data Networks. In: 8th Int. Conf. on Harmonisation within Atmospheric Dispersion Modelling for Regulatory Purposes (2002)

26. Wahab, A.-S.A., Al-Alawi, S.M.: Assessment and prediction of tropospheric ozone concentration levels using artificial neural networks. Environmental Modeling & Software 17, 219–228 (2002)

27. Wolpert, D.: Stacked Generalization. Neural Networks 5, 241–259 (1992)

28. Zhou, Z.H., Wu, J., Wei, T.: Corrigendum to "Ensembling neural networks: Many could be better than all". Artificial Intelligence 174(18), 15–70 (2010)

29. Gardner, M.W., Dorling, S.R.: Artificial Neural Networks (The Multilayer Perceptron) - a Review of Applications in the Atmospheric Sciences. Atmospheric Environment 32(14/15), 2627–2636 (1998)

30. Kolehmainen, M., Martikainen, H., Ruuskanen, J.: Neural networks and periodic components used in air quality forecasting. Atmospheric Environment 35(5), 815–825 (2001)

Web References

1. http://www.epa.gov/pm/
2. http://www.ypeka.gr/

Remarks on Computational Facial Expression Recognition from HOG Features Using Quaternion Multi-layer Neural Network

Kazuhiko Takahashi[1], Sae Takahashi[1], Yunduan Cui[2],
and Masafumi Hashimoto[3]

[1]Information Systems Design, Doshisha University, Kyoto, Japan
{katakaha@mail,buj1078@mail4}.doshisha.ac.jp
[2]Graduate School of Doshisha University, Kyoto, Japan
dum3101@mail4.doshisha.ac.jp
[3]Intelligent Information Engineering and Science, Doshisha University, Kyoto, Japan
mhashimo@mail.doshisha.ac.jp

Abstract. Facial expression recognition is an important technology in human-computer interaction. This study investigates a method for facial expression recognition using quaternion neural networks. A multi-layer quaternion neural network that conducts its learning using a quaternion back-propagation algorithm is employed to design the facial expression recognition system. The input feature vector of the recognition system is composed of histograms of oriented gradients calculated from an input facial expression image, and the output vector of the quaternion neural network indicates the class of facial expressions such as happiness, anger, sadness, fear, disgust, surprise and neutral. Computational experimental results show the feasibility of the proposed method for recognising human facial expressions.

Keywords: Quaternion neural network, Facial expression recognition, Histograms of oriented gradients, Image processing.

1 Introduction

Facial expression recognition is an interesting but difficult task, as facial expressions vary with age, race, gender, culture and so on. Because facial expression recognition is an important technology in human-computer interaction involving many fields such as image processing, pattern recognition, physiology and psychology, many studies have been conducted on this technology [1] [2] [3] [4]. One of the techniques employed for facial expression recognition involves the use of artificial neural networks because of their attractive features such as parallel distributed processing, learning, fault tolerance and robustness. Although conventional neural networks conduct signal processing involving real numbers, hyper-complex-valued neural networks that are based on Clifford algebra [5] (such as complex neural networks and quaternion neural networks) have been proposed [6] [7] [8] to solve classically hard-to-treat intractable problems by using

V. Mladenov et al. (Eds.): EANN 2014, CCIS 459, pp. 15–24, 2014.

real-valued neural networks. There have been many successful examples involving the use of such neural networks in applications requiring multi-dimensional signal processing, e.g. colour image processing [9], chaotic time-series prediction [10], multi dimensional time-series signal processing [11], inverse problem [12] and control of robot manipulator [13].

In this study, we investigate an approach to facial expression recognition using quaternion neural networks. Quaternion neural networks have been demonstrated better performances than real number neural networks because the quaternion neural network is able to cope with multidimensional issues more efficiently by employing quaternion directly. The histograms of oriented gradients (HOG) [14] [15] of the facial expression images are calculated so as to consist of the input feature vector of the quaternion neural network. In computational experiments of facial expression recognition, the quaternion neural network is trained and tested using image datasets that contain seven facial expressions (happiness, anger, sadness, fear, disgust, surprise and neutral). The feasibility of the quaternion neural network for this task was indicated by experimental results.

2   Facial Expression Recognition System

Figure 1 illustrates the facial expression recognition system using the quaternion neural network.

2.1   HOG Feature

HOG is a feature descriptor that describes local object appearance and shape within an image using the distribution of intensity gradients or edge directions. The procedure used to define the HOG feature is as follows: 1) calculating gradient directions, 2) compiling a histogram of gradient directions for the pixels within each cell, which divides the image into small connected regions and 3) normalising all cells within each block, which is a larger region of the image.

1) Gradient Computation

The gradient magnitude $a_{i,j}$ and direction $\theta_{i,j}$ at the pixel position $[i, i]$ in the image can be obtained by the following:

$$a_{i,j} = \sqrt{(I_{i+1,j} - I_{i-1,j})^2 + (I_{i,j+1} - I_{i,j-1})^2}, \tag{1}$$

$$\theta_{i,j} = \tan^{-1} \frac{I_{i,j+1} - I_{i,j-1}}{I_{i+1,j} - I_{i-1,j}}, \tag{2}$$

where $I_{i,j}$ is the intensity of the image at the pixel position $[i, j]$. To achieve better invariance to small variations in the image, the direction is constrained within the range $[0, \pi]$ by adding π to $\theta_{i,j}$ when the direction is negative.

2) Histogram Calculation

First, the image used to calculate the gradients is divided into small spatial regions called cells, with size $C_W \times C_H$. Then, a local one-dimensional histogram of gradient directions $\theta_{i,j}$ weighted by the gradient magnitude $a_{i,j}$ over the pixels of the cells is counted in each cell.

3) Normalisation

By grouping the cells together into larger spatially connected regions called blocks, with size $B_W \times B_H$, the local histogram is normalised over all the cells in the block as follows:

$$\bar{h}_{p_s,q_s} = \frac{h_{p_s,q_s}}{\sqrt{\|h_s\|^2 + e_0}}, \tag{3}$$

where h_{p_s,q_s} is the local histogram at the cell position $[p_s, q_s]$ in the s-th block, the element of vector h_s represents all local histograms over the cells in the s-th block and e_0 is an arbitrary small constant, which ensures that the denominator is not zero.

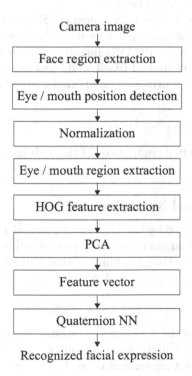

Fig. 1. Processing flow of facial expression recognition system

2.2   Feature Extraction from Face Image

Pretreatment of the input image using image processing is necessary before extracting the HOG features. First, the face region is found in the input image by a pattern classification method that utilises Haar-like feature templates. Eye and mouth positions are found using the same method, and their two-dimensional coordinates in the image are calculated. Next, the image is rotated such that both eye positions are on the horizon, and is scaled such that the distance between the eyes is a constant value D_e. In addition, the image is scaled to set the distance between the eye position and mouth position to a constant value D_m. Then, the face image with horizontal centre located between the eyes and vertical centre located at the mid-point between the eye and mouth is extracted from the input image with a size of $\alpha D_e \times \beta D_m$, where the coefficients α and β are determined based on the human body size database [16]. Finally, the eye region image with its centre located at the eye position is extracted from the face image and has a size of $W_e \times H_e$. In the same manner, the mouth region image with its centre is located at the mouth position is extracted from the face image and has a size of $W_m \times H_m$. The HOG features are calculated for the eye and mouth region images, respectively.

The HOG feature has a very high dimensional space. In order to compose feature vectors that are used as input vectors of the quaternion neural network, principal component analysis (PCA) is introduced to reduce the dimensions of the HOG features. PCA is applied to all HOG features obtained from the k-th image in each block, and r principal components are utilised to consist of the feature vector $\boldsymbol{v}_k$:

$$\boldsymbol{v}_k = \begin{bmatrix} {}^1v_1 \ {}^2v_1 \cdots {}^rv_1 \ {}^1v_2 \ {}^2v_2 \cdots {}^rv_2 \cdots\cdots {}^1v_N \ {}^2v_N \cdots {}^rv_N \end{bmatrix}^{\mathrm{T}}, \qquad (4)$$

where ${}^iv_j$ is the i-th principal component of the j-th block in the k-th image.

2.3   Quaternion Neural Network

The quaternion was invented by the Irish mathematician W. R. Hamilton in order to generalize complex number properties to multi-dimensional space. Quaternion forms a class of hyper complex number that consists of a real number and three imaginary numbers: i, j and k. A quaternion $\boldsymbol{q}$ is defined by:

$$\boldsymbol{q} = q_0 + q_1 \mathrm{i} + q_2 \mathrm{j} + q_3 \mathrm{k} = \begin{bmatrix} q_0 \ q_1 \ q_2 \ q_3 \end{bmatrix}^{\mathrm{T}}, \qquad (5)$$

where $q_i \ (i = 0, 1, 2, 3)$ is the real number parameter. The real number unit is 1 and the three imaginary units are i, j and k. They are orthogonal spatial vectors. The conjugate of a quaternion $\boldsymbol{q}^*$ is defined by:

$$\boldsymbol{q}^* = q_0 - q_1 \mathrm{i} - q_2 \mathrm{j} - q_3 \mathrm{k}, \qquad (6)$$

and the multiplication between one quaternion and its conjugate as follows:

$$\boldsymbol{q} \otimes \boldsymbol{q} = q_0^2 + q_1^2 + q_2^2 + q_3^2. \qquad (7)$$

Addiction and subtraction of two quaternions, $\boldsymbol{q}$ and $\boldsymbol{r}$, are defined by:

$$\boldsymbol{q} \pm \boldsymbol{r} = \begin{bmatrix} q_0 \pm r_0 & q_1 \pm r_1 & q_2 \pm r_2 & q_3 \pm r_3 \end{bmatrix}^{\mathrm{T}}. \tag{8}$$

Multiplication between a real number a and a quaternion $\boldsymbol{q}$ is given by:

$$a\boldsymbol{q} = aq_0 + aq_1\mathrm{i} + aq_2\mathrm{j} + aq_3\mathrm{k}, \tag{9}$$

while the multiplication between two quaternions $\boldsymbol{q}$ and $\boldsymbol{r}$ is given by:

$$\boldsymbol{q} \otimes \boldsymbol{r} = q_0 r_0 - \vec{q} \cdot \vec{r} + r_0 \vec{q} + q_0 \vec{r} + \vec{q} \times \vec{r} \tag{10}$$

where $\vec{q} = \begin{bmatrix} q_1 & q_2 & q_3 \end{bmatrix}^{\mathrm{T}}$, $\vec{r} = \begin{bmatrix} r_1 & r_2 & r_3 \end{bmatrix}^{\mathrm{T}}$, $\cdot$ and $\times$ represent scalar and vector product respectively. The norm of quaternion is defined by:

$$|\boldsymbol{q}| = \sqrt{\boldsymbol{q} \otimes \boldsymbol{q}^*}. \tag{11}$$

To describe the input and output relationships of the multi-layer quaternion neural network and the back-propagation algorithm, a three-layer quaternion neural network was considered. In the input layer of the quaternion neural network, the l-th neuron' input $\boldsymbol{x}_l$ is a quaternion:

$$\boldsymbol{x}_l = x_{0_l} + x_{1_l}\mathrm{i} + x_{2_l}\mathrm{j} + x_{3_l}\mathrm{k}, \tag{12}$$

In the hidden layer, the output from the m-th neuron unit $\boldsymbol{u}_m$ is defined as follows:

$$\boldsymbol{u}_m = f\left(\sum_l \boldsymbol{w}_{1_{ml}} \otimes \boldsymbol{x}_l + \boldsymbol{\phi}_{1_m}\right), \tag{13}$$

where $\boldsymbol{w}_{1_{ml}}$ is the weight between the l-th neuron of the input layer and the m-th neuron of the hidden layer, $\boldsymbol{\phi}_{1_m}$ is the threshold of the m-th neuron in the hidden layer, $f(\cdot)$ is an activation function which is split as follows:

$$f(\boldsymbol{x}) = f_0(\boldsymbol{x}_0) + f_1(x_1)\mathrm{i} + f_2(x_2)\mathrm{j} + f_3(x_3)\mathrm{k}, \tag{14}$$

In the output layer, the output from the n-th neuron unit $\boldsymbol{y}_n$ is defined by

$$\boldsymbol{y}_n = f\left(\sum_m \boldsymbol{w}_{2_{nm}} \otimes \boldsymbol{u}_m + \boldsymbol{\phi}_{2_n}\right), \tag{15}$$

where $\boldsymbol{w}_{2_{nm}}$ is the weight between the m-th neuron of the hidden layer and the n-th neuron of the output layer, and $\boldsymbol{\phi}_{2_n}$ is the threshold of the n-th neuron in the output layer.

The training of the quaternion neural network was carried out using the back-propagation algorithm to minimise the cost function J as follows:

$$\boldsymbol{w}_{2_{nm}}(t+1) = \boldsymbol{w}_{2_{nm}}(t) + \eta \boldsymbol{\delta}_{2_n}(t) \otimes \boldsymbol{u}_m^*, \tag{16}$$

$$\boldsymbol{\phi}_{2_n}(t+1) = \boldsymbol{\phi}_{2_n}(t) + \eta \boldsymbol{\delta}_{2_n}(t), \tag{17}$$

$$\delta_{2_n}(t) = \epsilon_n \odot f'\left(\sum_m w_{2_{nm}}(t) \otimes u_m + \phi_{2_n}\right), \tag{18}$$

$$w_{1_{ml}}(t+1) = w_{1_{ml}}(t) + \eta \delta_{1_m}(t) \otimes x_l^*, \tag{19}$$

$$\phi_{1_m}(t+1) = \phi_{1_m}(t) + \eta \delta_{1_m}(t), \tag{20}$$

$$\delta_{1_m}(t) = \left\{\sum_n w_{2_{nm}}(t) \otimes \delta_{2_n}(t)\right\} \odot f'\left(\sum_l w_{1_{ml}} \otimes x_l + \phi_{1_m}\right), \tag{21}$$

$$J = \frac{1}{2}\sum_n \epsilon_n \otimes \epsilon_n^*, \tag{22}$$

where η is the learning factor, ϵ_n is the output error defined by $\epsilon_n = d_n - y_n$, d_n is the desired output of the n-th neuron in the output layer and $\odot$ denotes the component-by-component product.

3   Computational Experiment for Recognising Facial Expression

In the experiment for the computational recognition of facial expression, six facial expressions that have been proposed as basic facial expressions by P. Ekman, together with one neutral facial expression were considered.

The quaternion neural network was trained and tested using facial expression images recorded from one Japanese male. Each facial expression was consciously

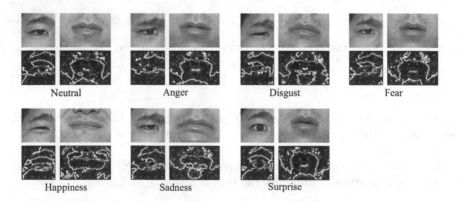

Fig. 2. Examples of input image and HOG feature seen with the facial expressions (top: input image, bottom: HOG feature, left: eye region, right: mouth region)

Table 1. Confusion matrix of computational facial expression recognition using quaternion neural network of 9–2–2 network topology

	neutral	anger	disgust	fear	happiness	sadness	surprise
neutral	**0.330**	0.025	0.065	0.225	0.035	0.195	0.125
anger	0.095	**0.575**	0.025	0.050	0.085	0.170	0.000
disgust	0.185	0.025	**0.455**	0.135	0.070	0.070	0.060
fear	0.450	0.015	0.080	**0.365**	0.015	0.075	0.000
happiness	0.005	0.070	0.060	0.025	**0.695**	0.110	0.035
sadness	0.165	0.165	0.095	0.085	0.080	**0.380**	0.030
surprise	0.200	0.000	0.015	0.125	0.005	0.025	**0.630**

Table 2. Confusion matrix of computational facial expression recognition using conventional real-valued neural network of 34–7–7 network topology

	neutral	anger	disgust	fear	happiness	sadness	surprise
neutral	**0.165**	0.005	0.035	0.145	0.075	0.265	0.310
anger	0.050	**0.290**	0.030	0.060	0.095	0.260	0.215
disgust	0.085	0.000	**0.265**	0.145	0.040	0.255	0.210
fear	0.165	0.005	0.030	**0.235**	0.050	0.270	0.245
happiness	0.045	0.010	0.010	0.060	**0.350**	0.250	0.275
sadness	0.130	0.000	0.025	0.125	0.070	**0.480**	0.170
surprise	0.115	0.000	0.025	0.095	0.025	0.040	**0.700**

expressed by the subject according to psychological knowledge [17], and 10 images were acquired for each facial expression. Thus, 70 image samples together with the label of the facial expression were obtained. Figure 2 shows an example of the input image and HOG feature, where the size of the eye and mouth regions is 60[pixel] × 60[pixel] and 90[pixel] × 60[pixel], respectively. While calculating the HOG feature, the cell size of the eye region was 12[pixel] × 12[pixel], that of the mouth region was 12[pixel] × 12[pixel] and the block size for both regions was 3 × 3. The feature vector v was composed with the first and second principal components of the PCA results for all HOG features. As a result, the dimension of the feature vector in the eye region is 18, and that in the mouth region is 16. The dimension of the input vector of the quaternion neural network was 34 because it consisted of the eye and mouth region feature vectors. The topology of the quaternion neural network was a 9–2–2 network (104 parameters), and a sigmoid function was used as the function $f_r(\cdot)$ $(r = 0, 1, 2, 3)$ in the hidden and output layers. The output from the quaternion neural network represents the likelihood that the input feature vector corresponds to the facial expressions. The quaternion neural network's output that gives the largest value is chosen, and the corresponding class indicates the recognition result. Training and testing of the quaternion neural network for facial expression recognition were conducted by leave-one-out cross-validation with 20 different initial weight conditions. Table 1 shows the results of the computational facial expression recognition, and

Table 3. Confusion matrix of person-dependent computational facial expression recognition using JAFFE datasets

	neutral	anger	disgust	fear	happiness	sadness	surprise
neutral	**0.586**	0.138	0	0.034	0.103	0.034	0.103
anger	0.032	**0.452**	0.161	0.194	0.032	0.097	0.032
disgust	0.034	0.069	**0.414**	0.310	0.103	0.069	0
fear	0.186	0.094	0.156	**0.375**	0.094	0.031	0.063
happiness	0.129	0.129	0	0.065	**0.484**	0.032	0.161
sadness	0.167	0.133	0.067	0.167	0.033	**0.430**	0.033
surprise	0.033	0.100	0.033	0.100	0.033	0.033	**0.677**

Table 4. Confusion matrix of person-independent computational facial expression recognition using JAFFE datasets

	neurtal	anger	disgust	fear	happiness	sadness	surprise
neutral	**0.183**	0.137	0.237	0.147	0.1	0.077	0.090
anger	0.137	**0.217**	0.163	0.120	0.080	0.143	0.110
disgust	0.190	0.162	**0.114**	0.134	0.131	0.131	0.107
fear	0.144	0.138	0.184	**0.206**	0.069	0.125	0.094
happiness	0.158	0.126	0.100	0.110	**0.216**	0.119	0.148
sadness	0.100	0.155	0.152	0.171	0.110	**0.135**	0.132
surprise	0.137	0.220	0.120	0.130	0.117	0.130	**0.133**

the averaged recognition rate was 49%; however, there remains some difficulty in recognising negative facial expressions such as sadness, fear and neutral. As a reference for comparing the result of computational facial expression recognition, conventional real-valued neural network was utilized for this task. The topology of neural network was a $34 - M - 7$ network in which the sigmoid function was used in the hidden layer and the linear function was utilized in the input and output layers. Training of the neural network was carried out by using the back-propagation algorithm. The averaged recognition rates are 30% ($M = 2$, 91 parameters), 35% ($M = 7$, 301 parameters) and 34% ($M = 10$, 427 parameters) and Table 2 shows the results of the computational facial expression recognition using a 34-7-7 network topology. This result shows that the quaternion neural network can provide a better performance than the conventional real-valued neural network with fewer number of parameters.

Next, the quaternion neural network was trained and tested using the Japanese female facial expression datasets [18]. This database contains 213 images of seven facial expressions shown by 10 Japanese females. The HOG feature was calculated with the cell size of 8[pixel] × 8[pixel] in the eye region of 48[pixel] × 48[pixel] and that of 6[pixel] × 8[pixel] in the mouth region of 60[pixel] × 40[pixel]. The block size for both regions was 3 × 3. Applying the PCA to each HOG feature, the dimension of the input vector of the quaternion neural network was 80. The topology of the quaternion neural network was a 20–10–5–2 network. Training and testing of the quaternion neural network were conducted by leave-one-out cross-validation with 10 different initial weight conditions. Tables 3 and 4 show the results of the computational facial expression recognition. Table 3 shows the average of the person-dependent recognition result where the training and testing of the quaternion neural network were conducted with the images of the target subject. Table 4 shows the average of the person-independent recognition where the training of the quaternion neural network was conducted except for the images of the target subject and then the testing of the quaternion neural network was carried out by the images of the target subject. The averaged recognition rate of 50% was achieved in person-dependent facial expression recognition, however, that of 17% was attained in person independent facial expression recognition.

These experimental results show the potential of achieving computational facial expression recognition by using the quaternion neural network with the HOG feature, however the recognition rate using the quaternion neural network is not sufficient. Extraction of possible features from facial images in addition to the HOG feature should be investigated to improve the recognition ability.

4   Conclusions

This study investigated computational facial expression recognition using a multi-layer quaternion neural network that conducted its learning using the quaternion back-propagation algorithm. The proposed recognition process involved face image pre-treatments using image processing, feature extraction from images using the HOG calculation, dimension reduction of the feature vector using PCA and the evaluation of outputs from the quaternion neural network. Face image datasets that contained seven facial expressions, namely happiness, anger, sadness, fear, disgust, surprise and neutral, were used to evaluate the recognition method. In computational experiments performed to recognise facial expressions, the quaternion neural network attained an averaged recognition rate of around 50% for all seven facial expressions in person-dependent recognition experiments. The results obtained in this study demonstrated that the quaternion neural network is feasible for computational facial expression recognition.

Acknowledgements. We would especially like to thank Mr. Kohei Morii for his help in this work.

References

1. Tolba, A.S., El-Baz, A.H., El-Harby, A.A.: Face Recognition: A Literature Review. International Journal of Information and Communication Engineering 2, 88–103 (2006)
2. Zeng, Z., Pantic, M., Roisman, G.I., Huang, T.S.: A Survey of Affect Recognition Methods: Audio, Visual, and Spontaneous Expressions. IEEE Transactions on Pattern Analysis and Machine Intelligence 31(1), 39–58 (2009)

3. Calvo, R.A., D'Mello, S.: Affect Detection: An Interdisciplinary Review of Models, Methods, and Their Applications. IEEE Transactions on Affective Computing 1(1), 18–37 (2010)
4. Sandbach, G., Zafeiriou, S., Pantic, M., Yin, L.: Static and Dynamic 3D Facial Expression Recognition: A Comprehensive Survey. Image and Vision Computing 30, 683–697 (2012)
5. Sommer, G. (ed.): Geometric Computing with Clifford Algebra. Springer (2001)
6. Buchholz, A., Sommer, G.: On Clifford Neurons and Clifford Multi-Layer Perceptrons. Neural Networks 21, 925–935 (2008)
7. Nitta, T. (ed.): Complex-Valued Neural Networks: Utilizing High-Dimensional Parameters. Information Science Publishing (2009)
8. Kuroe, Y.: Models of Clifford Recurrent Neural Networks and Their Dynamics. In: Proceedings of International Joint Conference on Neural Networks, pp. 1035–1041 (2011)
9. Kusamichi, H., Isokawa, T., Matsui, N., Ogawa, Y., Maeda, K.: A New Scheme Color Night Vision by Quaternion Neural Network. In: Proceedings of the 2nd International Conference on Autonomous Robots and Agents, pp. 101–106 (2004)
10. Arena, P., Caponetto, R., Fortuna, L., Muscato, G., Xibilia, M.G.: Quaternionic Multilayer Perceptrons for Chaotic Time Series Prediction. IEICE Transactions on Fundamentals E79-A(10), 1682–1688 (1996)
11. Ujang, B.C., Took, C.C., Mandic, D.P.: Quaternion-Valued Nonlinear Adaptive Filtering. IEEE Transactions on Neural Networks 22(8), 1193–1206 (2011)
12. Iura, T., Ogawa, T.: Quaternion Neural Network Inversion for Solving Inverse Problems. In: Proceedings of SICE Annual Conference 2012, pp. 1802–1805 (2012)
13. Cui, Y., Takahashi, K., Hashimoto, M.: Remarks on Quaternion Neural Networks with Application to Robot Control. In: Proceedings of SICE 2013 International Conference on Instrumentation, Control, Information Technology and System Integration, pp. 1381–1386 (2013)
14. Dalal, N., Triggs, B.: Histograms of Oriented Gradients for Human Detection. In: Proceedings of the 2005 IEEE Computer Society Conference on Computer Vision and Pattern Recognition, pp. 886–893 (2005)
15. Onishi, K., Takiguchi, T., Ariki, Y.: 3D Human Posture Estimation Using the HOG Features from Monocular Image. In: Proceedings of 19th International Conference on Pattern Recognition, pp. 1–4 (2008)
16. http://riodb.ibase.aist.go.jp/dhbodydb/index.php.ja
17. Ekman, P., Friesen, W.V.: Unmasking The Face. Prentice-Hall, Inc., Englewood Cliffs (1975)
18. Lyons, M.J., Akamatsu, S., Kamachi, M., Gyoba, J.: Coding Facial Expressions with Gabor Wavelets. In: Proceedings of 3rd IEEE International Conference on Automatic Face and Gesture Recognition, pp. 200–205 (1998)

Classification of Database by Using Parallelization of Algorithms Third Generation in a GPU

Israel Tabarez Paz[1], Neil Hernández Gress[2], and Miguel González Mendoza[2]

[1] Universidad Autónoma del Estado de México
Blvd. Universitario s/n, Predio San Javier, Atizapán de Zaragoza, México
itabarezp@uaemex.mx
http://www.uaem.mx/cuyuaps/vallemexico.html
[2] Tecnológico de Monterrey, Campus Estado de México,
Carretera Lago de Guadalupe km 3.5, Col. Margarita Maza de Juarez,
Atizapán de Zaragora, México
{ngress,mgonza}@itesm.mx
http://www.itesm.edu

Abstract. This manuscript is focused on the efficiency analysis of Artificial Neural Networks (ANN) that belongs to the third generation, which are Spiking Neural Networks (SNN) and Support Vector Machine (SVM). The main issue of scientific community have been to improve the efficiency of ANN. So, we applied architecture GPU (Graphical Processing Unit) from NVIDIA model GeForce 9400M. On the other hand, the results of QP method for SVM depends on computational complexity of the algorithm, which is proportional to the volume and attributes of the data. Moreover, SNN was selected because it is a method that has not been explored fully. Despite the economic cost is very high in parallel programming, this is compensated with the large number of real applications such as clustering and pattern recognition. In the state of the art, nobody of authors has coded Quadratic Programming (QP) of SVM in a GPU. In case of SNN, it has been developed by using a specific software as MATLAB, FPGA or sequential circuits but it have never been coded in a GPU. Finally, it is necessary to reduce the grade of parallelization caused by limitations of hardware.

Keywords: GPU, CPU, Artificial Neural Networks, Spiking Neural Networks, Support Vector Machine, Classification.

1 Introduction

In this paper presents the analysis and comparison of efficiency between methodologies Quadratic Programming (QP) of Support Vector Machine (SVM) and Spikeprop of Spiking Neural Networks (SNN) by using parallel programming in a GPU.

Many authors in the area of Artificial Neural Networks (ANN) are interested on thee methodologies in order to improve the efficiency according to the right

V. Mladenov et al. (Eds.): EANN 2014, CCIS 459, pp. 25–38, 2014.

algorithm for each application. In general, there are several methods, the most used are decision trees, Bayesian networks, other neural networks, statistical methods, genetic algorithms, fuzzy logic, Markov's model, etcetera. Additionally, there are others as Sequential Minimal Optimization (SMO) [23] used for Support Vector Machine (SVM) [8]; and Izhikevich's model [12] and Hodgkin-Huxley's model [11], for Spiking Neural Networks (SNN) they both. In this case, our interest in because QP of SVM finds the Global Minimum Value, but its computational complexity is very high. Also, we are interested on studying the characteristics and behavior about learning time of SNN, that has capacity of working at the same iteration with multiclass data. However, an important disadvantage is that SNN only finds the local minimum value.

On the other hand, we compared the efficiency between algorithms QP from SVM and Spikeprop from SNN. The result allows us to select the right methodology for each application and to know the pending tasks. According to application, SVM algorithm has been used for clustering, classification, pattern recognition and regression. For example, Carpenter [6] use LIBSVM of MATLab for classification and regression in order to compare results with a known software called CUSVM [6], what is used for parallel programming, and applied a version of Sequential Minimal Optimization (SMO). Also, Markos Papadonikolakis [21] was focused on speeding up for classification by using a GPU and a FPGA (Field Programmable Gate Array). Other author, Tsung-Kai Lin and Shao-Yi Chien [16] continued with the works of Catanzaro [7] and Carpenter [6] for speeding up the SVM algorithm. He applied an algorithm of Sparse Matrix Multiplication in order to solve problems of classification.

In the case of algorithm SNN, there are not many published applications, but Thomas Nowotny [20] implemented a realistic morphology of SNN of smell of insects. In case of Bhuiyan [2], he programmed the Izhikevich's model [12] for character recognition. Sander [4] did implement on a GPU the algorithm of SNN but he only proposes an application of clustering. Also, OlafOlaf applied SNN for voice recognition. However, we also suggest some applications about characters recognition considering only the five vowels, but in this problem we can use more letters and numbers. The result allows us to define some issues in the future. So, we used a real database about patients in a hospital for classification according to some proposed characteristics.

The parallel programming [19] consists on solving big problems by parts. However, there are three forms of parallel programming: bit level, instruction level, and task parallelism. The selected architecture and applications are oriented to instruction level by using a GPU. Problems of parallel programming are solved in language C for CUDA.

This paper is distributed as follows: in section 2 related works are presented, in section 3 the parallelization of the algorithm SNN is presented, in section 4 the parallelization of the algorithm SVM is presented, in section 5 correspond to the experiments of SNN and SVM, in section 6 the experiments are analyzed and presented, finally in section 7 we present the conclusion and future work.

2   Related Works

In this section we present the works related about this topic. The figure 1 describes chronologically the evolution about state of the art of SNN. We only show the main authors.

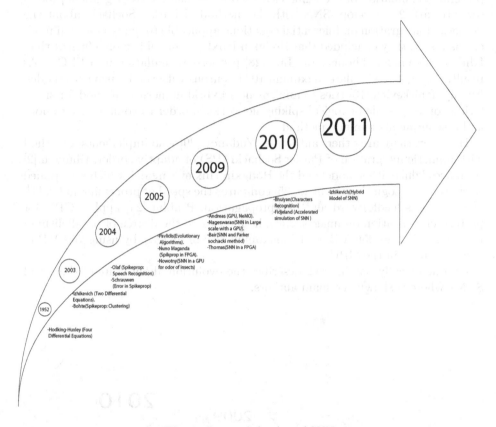

Fig. 1. State of the Art of SNN

In the last decade, evolution of the SNN model has been increased very fast because some authors have focused on type of hardware for programming and its applications.Firstly, Hodking-Huxley [11] sets the bases about how SNN operates in mammal brain. He proposes a mathematical model with four differential equations. In middle of century XIX, Izhikevich [12] proposes a simpler model that Hodking-Huxley, only with two differential equations. Also Bohte [3] modeled the algorithm of SNN with version of the algorithm of Backpropagation [14] for applications of clustering. A year later, Olaf [5] continues at the same line of Bohte but he proposes an application of Speech Recognitions, which it was restricted only for the numbers one and two. In this case, as future work, we can extend the experiment to more numbers. Schrauwen [26] proposes a way to improve the algorithm of Spikeprop of SNN. Pavlidis [22] implements SNN with evolutionary algorithms. Maganda [18] programmed the algorithm Spike-prop from SNN in a FPGA. In contrast, Nowotny [20] programmed the algorithm in a GPU. Also, he applies the SNN algorithm for simulating olfatory of insects. Andreas [9] compares the execution of SNN of a GPU with a special card called NeMO, however this card is not very commercial. Nageswaran [17] presents a simulation of the Izhikevich's model by using CUDA graphic processor. Stewart [27] develops SNN with the method of Parker Sochacki about the numerical integration of differential equations applicable to many neuronal models, as result, they concluded that Hodgkin Huxley's model is more efficient than Izhikevich's model. Thomas and Luk [28] propose a simulation in a FPGA. As result, they got to simulate maximum 1024 neurons with the Izhikevich's model. Finally, Izhickevich [13] tries to implement a hybrid numerical method for simulations of large-scale biological spiking networks in order to combine continuous and discontinuous numeric methods.

However, there are others authors as Yudanov [29] who implements a method with numeric integration of Parker Sochacki (PS) of adaptive order. Bhuiyan [2] compares Izhikevich's model and the Hodgkin Huxley's model, which are applied to character recognition. Scanzio [25] compares the speed of processing in CUDA of algorithms feedforward and backpropagation. Prabhu [24] applies GPU for pattern classification on images. Also, he focuses on the degree of parallelism of problems. He uses 256 MB as the maximum size of images, by using 768 MB of video memory in the GPU.

Chronologically, the figure 2 describes the evolution about state of the art of SVMs where is shown the main authors.

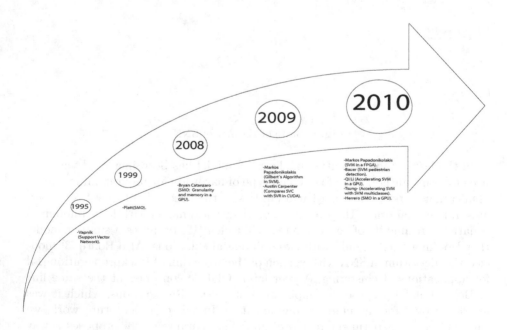

Fig. 2. State of the Art of SVM

On the other side, Support Vector Machine were developed in 1995 by Vapnick [8], by using the method Quadratic Optimization (QP). In spite of SVM finds the Global Minimum Value, they spends a lot of computational memory. However, in 1999 Platt [23] proposes the method Sequential Minimal Optimization (SMO) applieed to SVM, which divides all problem in subsets, but it only finds the local minimum value. Also, Catanzaro [7] optimized SVM in the method Sequential Minimal Optimization (SMO) by using a GPU. In this case, the Learning Time and precision between the GPU and libraries about SVM of MatLab are compared. Papadonikolakis [21] focused on SVM by using the Gilbert's algorithm on a FPGA in order to compare the speed of learning and efficiency. In the same researcher line Carpenter [6] applies cuSVM for NVIDIA with a modified version of SMO in order to compare SVC with SVR. After that, Bauer [1] applies GPU for detection of pedestrians. Qi Li [15] applies SVM by using a GPU for classification problems. Also, Tsung [16] implements the methodology SVM in a GPU. Finally, Herrero [10] made a classification of database by using SVM in a GPU, he continues Catanzaro's work.

In the next section we explain about parallelization of algorithms SNN and SVM.

3   Parallelization of SNN

In this paper we propose an architecture of three layers i, j y k, because of limitation computational resources, although it can be expanded for more layers and neurons. So, it is necessary to expand the quantity of blocks in GPU for bigger databases, what implies economical cost is higher. In the figure 3 is described the architecture of Spiking Neural Networks for three layers with p neurons of input, q neurons of the hidden layer and one neuron in the output layer.

The synaptic delays are defined as neural connections from one layer to another. In the figure 4 the delays are described by the synaptic connections. According to Olaf [5] the interval of delays depend on the input data and the desired data. So, this author suggest calculating the smallest delays as the difference between the last spike from the input to the earliest spike in the output, also the largest delay is the difference between the time from the earliest spike in the input to the last spike in the output. Every spike in the input can be delayed a degree, this can influence in the early and late desired spike in the output.

A neuron j, belongs to a set Γ_j (presynaptic neuron) [5]. The neuron is fired when it receives a set of spikes encoded on the time $t_i, i \epsilon \Gamma_j$ whose sum reaches the threshold θ. The variable $x_j(t)$ is given by the following equation (1) where w_{ij} are the weights from the connection i to j:

$$x_j(t) = \sum_{i \epsilon \Gamma_j} w_{ij} \varepsilon(t - t_i) \tag{1}$$

Also, $\varepsilon(t)$ is the synaptic potential described in the equation (2)

$$\varepsilon(t) = \frac{t}{\tau} e^{1 - \frac{t}{\tau}} \tag{2}$$

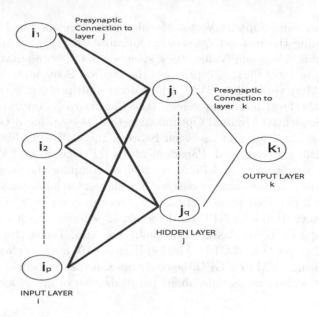

Fig. 3. Delays as Synaptic Connections

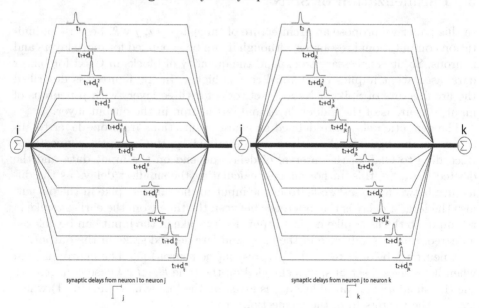

Fig. 4. Delays as Synaptic Connections

Equation (3) describes the quantity of synaptic conexions of a neuron j where τ is a constant that controls the width of the pulse.

$$x_j(t) = \sum_{i\epsilon\Gamma_j}\sum_{k=1}^{m} w_{ij}^k y^k(t) \tag{3}$$

The output of a neuron is described by 4:

$$y_j^k(t) = \varepsilon(t - t_i - d_k) \tag{4}$$

Where d_k is the delay time of a conection k fired of a presynaptic neuron. Finally, equation 5 represents the output potential $u_j(t)$ for a j neuron:

$$u_j(t) = \sum_{t_j^{(f)}\epsilon F_j} \eta(t - t_j^{(f)}) + \sum_{i\epsilon\Gamma_j}\sum_{t_i^{(g)}} w_{ij}\varepsilon(t - t_i^{(g)} - d_{ij}) \tag{5}$$

Where,

$$F_j = \{t^{(f)}; 1 \le f \le n\} = \{t|u_j(t) = \vartheta\} \tag{6}$$

The card for parallel programming has 255 threads per block, and 235 blocks, so the total quantity of threads in the card is 60000, aproximately. The equation 7 describes how the threads per blocks were calculated. Where $threadx$ is the quantity of threads calculated what represents the time t as the figure 4; $SIZE$ is threads per block; $NN[num\_cap]$ is the number of neurons in the layer; $NN[num\_cap + 1]$ is the number of neurons in the next layer. Delays are taken into account sequentially in the CPU (Central Processing Unit) because of limitations of memory.

$$threadx = \frac{SIZE}{(NN[num\_cap])(NN[num\_cap + 1])} \tag{7}$$

Neurons are parallelized as it is shown in the figure 5 where we can see a cubical matrix, which represents a block configured in three dimensions. The main parameters configured in the block of the GPU are the number of neurons in the current layer, the number of neurons in the next layer and the time.

4 Parallelization of SVM

According to optimization method applied for SMO, SVM light or QP, the parallelization of SVM is applied on operations of matrices, in contrast with SNN, the parallelization is focused on using of the architecture of the network and some mathematical operations. However, the problem on memory limitations is that our GPU can only store 256 data per block to be parallelized. So, this implies that the maximum number of data in the input are 16. As a consequence, we need to increse the quantity of blocks configured in the GPU to cover all elements of the database. However, when the amount of data invade more blocks, so the operations between blocks should be programmed sequentially because of

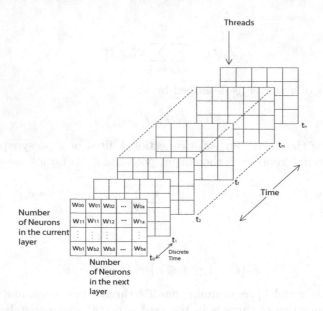

Fig. 5. Parallelization of SNN in one Block of the GPU configured in three dimensions

synchronization could be affected. If we work with the 60,000 threads availables in the GPU, then the quantity of input datas are 244. That implies to reduce the level of parallelization.

The mathematical model of QP from SVM is described as in equation 8:

$$\text{Maxime } \alpha$$
$$q(\boldsymbol{\alpha}) = 0.5\boldsymbol{\alpha}^T Q \boldsymbol{\alpha} - 1^T \alpha \tag{8}$$
$$\text{subject to}$$
$$\boldsymbol{y}^T \boldsymbol{\alpha} = 0$$
$$0 \leq \boldsymbol{\alpha} \leq \boldsymbol{C}$$

The matrix $\boldsymbol{Q}$ is composed by $(\boldsymbol{Q})_{ij} = y_i y_j k(x_i, x_j)$, where:
$i, j = 1, 2, 3, ..., l$, $\boldsymbol{\alpha} = [a^1...a^l]^T$
$1 = [1^1...1^l]^T$ $\boldsymbol{y} = [y^1...y^l]^T$
$\boldsymbol{C} = [C^1...C^l]^T$
The output is: $y = sign(\sum_{i=1}^{N} y_i \alpha_i K(\boldsymbol{x}, \boldsymbol{x_i}))$.
In this case, figure 6 presents the architecture of SVM.
For matrix multiplication for $((Q)_{ij} = y_i y_j k(x_i, x_j))$ the quantity of threads was generalized in the equation 9 and 10 :

$$BlocksPerRow = \frac{NUMBER\ OF\ DATAS}{THREADS\ PER\ BLOCK} + 1 \tag{9}$$

$$ThreadPerRow = \frac{NUMBER\ OF\ DATAS}{BlocksPerRow} \tag{10}$$

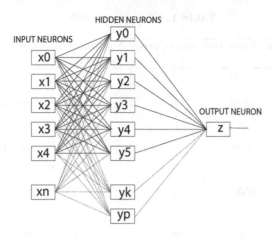

Fig. 6. Architecture of SVM

5   Experiments

5.1   Spiking Neural Network (SNN)

In this section the result obtained is shown of the parallelization of SNN. In the Table 1 we can see Learning Time for several database.

One problem of SNN, is about encoding the input to spike train. SNN use a sets of spikes as input and output. Analog values can be encoded sequentially during a certain period of time. Another simple way to encode a stream of analog values is by thresholding. Also, other way is fast Fourier transform used to convert sound samples into different frequency-signals [5]. It is difficult to say which method is good or not, but some seem more suited for certain applications than others. In our case, we took into accounto a time window for the logic output. In other words, if our desired output is 10ms, so our 1 logic is detected if the obtained output is from 9 to 11. In contrast, our 0 logic is detected if the obtained output is from 7 to 9.

5.2   Support Vector Machine (SVM)

As we have said, SVM are used for classification, regression, prediction and densities of data, although in this case it was used for classification. In case of SVM multiclass, we applied serial networks, it means that the output of the first neural network is the input of the next neural network. Where the number of serial networks is calculated as11:

$$(Number\ of\ serial\ networksl\ SVM) = (Classes) - 1 \qquad (11)$$

Respect to multiclass, is important to tell that the kernel of each class can not be found because the hyperplane in each class has different dimension, for example case of Database of Cars, each group was processed in different kernel: Iris was calculated with polynomial grade 2; the class 1 of Cars with "erbf 3",

Table 1. Efficiency of SNN

Database	Instances	Attributes	Classes	Number of Iterations	Successes	Learning Time [ms]
Iris	150	4	3	12	149	52331.625
Cars	1729	6	4	12	1294	1319680.625
Breast Cancer Wisconsin	699	9	2	19	673	253460.0937
Adult	32561	14	2	30	29247	24 hrs aprox.
Heart Disease Cleveland (HDC)	303	13	5	40	221	1069932.75

Table 2. Efficiency of SVM

Database	Instances	Attributes	Classes	Number of Iterations	Successes	Learning Time [ms]
Iris	150	4	3	2758	145	1521128.5
Cars (SVM light)	1728	6	4	——	1728 (Clase 1)	608927.125
					56 (Clase 2)	1478488.5
					4 (Clase 3)	1090742.125
					34 (Clase 4)	970055.375
Breast Cancer Wisconsin	699	9	2	——	641 (Clase 1)	163790.55625
					127 (Clase 2)	162216.453125
Adult	32561	14	2	——	——	——
Heart Disease Cleveland (HDC)	303	13	5	——	147 (Clase 1)	56749.73046
					3 (Clase 2)	21624.41406
					66 (Clase 3)	22715.71875
					25 (Clase 4)	27424.23046
					2 (Clase 5)	98908.94531

the class 2 with polynomial grade 3, the class 3 with polynomial grade 6 and the class 4 with polynomial grade 8.

The results for QP – SVM are shown in the Table 2.

6   Results

In this section results are compared between the two methodologies. The figure 7 shows the comparison of learning time and the figure 8 shows the comparison about quantity of successes per methodology.

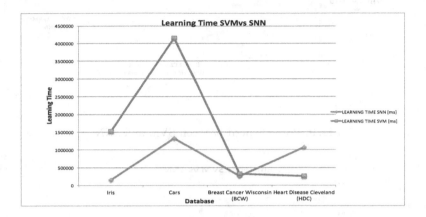

Fig. 7. Comparison of Learning Time between SNN and SVM

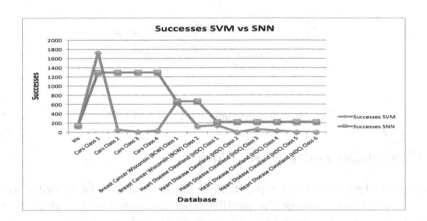

Fig. 8. Comparison of Successes between SNN and SVM

According to Table 2 we compare the rate of speed between SVM and SNN what is shown in the figure 9, in case of database Iris, SVM was slower than SNN, however for HDC SNN was 4.7 times slower than SVM with 80.198% of successes for SVM and 92.94% of successes for SNN.

In the figure 10, the comparison of efficiency between two the methodologies is shown. Efficiency was calculated by dividing the input data with output data. In this figure, we see that SNN is more efficient than SVM for databases with more of 150 instances.

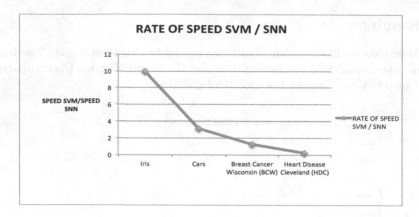

Fig. 9. Rate of speed between SVM and SNN

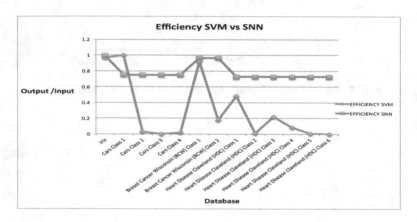

Fig. 10. Efficiency between SVM and SNN

7   Conclusion and Future Work

The viewed algorithms can be parallelized approximately 90% in order to increase its efficiency, however the parallelization is limited to the architecture, so in this case parallelization has to be sufficiently reduced. On the other side, the best solution of SNN depends on data encoding and the right values of the parameters of the neural network. Also, the speed depends on the hardware architecture. However, in spite of the efficiency of SNN, we proved that SVM finds the Global Minimum Value, in contrast SNN finds the Local Minimum Value.

In the future work is necessary to do the comparison by using a double precision hardware for getting more precision, so the efficiency will increase. Additionally, is important to develop a methodology for calculating the optimal values of the parameters of SNN in ordet to the solution quickly. Finally, as future work we suggest for researching of encoding the input data of SNN because speed depends on this.

References

1. Bauer, S., Kohler, S., Doll, K., Brunsmann, U.: Fpga-gpu architecture for kernel svm pedestrian detection. In: 2010 IEEE Computer Society Conference on Computer Vision and Pattern Recognition Workshops (CVPRW), pp. 61–68. IEEE (2010)
2. Bhuiyan, M.A., Pallipuram, V.K., Smith, M.C.: Acceleration of spiking neural networks in emerging multi-core and gpu architectures. In: 2010 IEEE International Symposium on Parallel & Distributed Processing, Workshops and Phd Forum (IPDPSW), pp. 1–8 (April 2010)
3. Bohte, S.M.: Spiking neural networks. Unpublished doctoral dissertation, Centre for Mathematics and Computer Science, Amsterdam (2003)
4. Bohte, S.M., Kok, J.N., La Poutre, H.: Error-backpropagation in temporally encoded networks of spiking neurons. Neurocomputing 48(1), 17–37 (2002)
5. Booij, O.: Temporal pattern classification using spiking neural networks. Unpublished master's thesis, University of Amsterdam (August 2004)
6. Carpenter, A.: Cusvm: A cuda implementation of support vector classification and regression (2009), patternsonscreen.net/cuSVMDesc.pdf
7. Catanzaro, B., Sundaram, N., Keutzer, K.: Fast support vector machine training and classification on graphics processors. In: Proceeedings of the 25th Intenational Conference on Machine Learning, pp. 104–111 (2008)
8. Cortes, C., Vapnik, V.: Support-vector networks. Machine Learning 20(3), 273–297 (1995)
9. Fidjeland, A.K., Roesch, E.B., Shanahan, M.P., Luk, W.: Nemo: A platform for neural modelling of spiking neurons using gpus. In: 20th IEEE International Conference on Application-specific Systems, Architectures and Processors, ASAP 2009, pp. 137–144 (July 2009)
10. Herrero-Lopez, S., Williams, J.R., Sanchez, A.: Parallel multiclass classification using svms on gpus. In: Proceedings of the 3rd Workshop on General-Purpose Computation on Graphics Processing Units, pp. 2–11. ACM (2010)
11. Hodgkin, A.L., Huxley, A.F.: A quantitative description of membrane current and its application to conduction and excitation in nerve. The Journal of physiology 117(4), 500 (1952)
12. Izhikevich, E.M.: Simple model of spiking neurons. IEEE Transactions on Neural Networks 14(6), 1569–1572 (2003)
13. Izhikevich, E.M.: Hybrid spiking models. Philosophical Transactions of the Royal Society A: Mathematical, Physical and Engineering Sciences 368(1930), 5061–5070 (2010)
14. Laurene, F.: Fundamentals of Neural Networks, Architecture, Algorithms, and Applications. Prentice Halls (1994)
15. Li, Q., Salman, R., Kecman, V.: An intelligent system for accelerating parallel svm classification problems on large datasets using gpu, pp. 1131–1135
16. Lin, T.-K., Chien, S.-Y.: Support vector machines on gpu with sparse matrix format. In: 2010 Ninth International Conference on Machine Learning and Applications (December 2010), pp. 313–318 (2010)
17. Nageswaran, J.M., Dutt, N., Krichmar, J.L., Nicolau, A., Veidenbaum, A.V.: Efficient simulation of large-scale spiking neural networks using cuda graphics processors. Neural Networks, 791–800 (June 2009)

18. Nuno-Maganda, M.A., Arias-Estrada, M.O.: Real-time fpga-based architecture for bicubic interpolation: an application for digital image scaling. In: International Conference on Reconfigurable Computing and FPGAs, ReConFig 2005, p. 8. IEEE (2005)
19. NVIDIA. NVIDIA CUDA C BEST PRACTICES GUIDE DG –05603–001v5.0, dg –05603–001v5.0 ed. (May 2012)
20. Nowotny, T., Huerta, R., Abarbanel, H.D., Rabinovich, M.I.: Self-organization in the olfactory system: one shot odor recognition in insects. Biological Cybernetics 93(6), 436–446 (2005)
21. Papadonikolakis, M., Bouganis, C.-S., Constantinides, G.: Performance comparison of gpu and fpga architectures for the svm training problem. In: International Conference on Field Programmable Technology, FPT 2009, pp. 388–391. IEEE (2009)
22. Pavlidis, N., Tasoulis, O., Plagianakos, V.P., Nikiforidis, G., Vrahatis, M.: Spiking neural network training using evolutionary algorithms. In: 2005 IEEE International Joint Conference on Neural Networks, IJCNN 2005, vol. 4, pp. 2190–2194. IEEE (August 2005)
23. Platt, J.C.: Sequiential minimal optimization: A fast algorithm for tarining support vector machines
24. Prabhu, R.D.: Somgpu: An unsupervised pattern classifier on graphical processing unit. In: IEEE Congress on Evolutionary Computation, CEC 2008 (IEEE World Congress on Computational Intelligence), pp. 1011–1018, 1–6 (June 2008)
25. Scanzio, S., Cumani, S., Gemello, R., Mana, F., Laface, P.: Parallel implementation of artificial neural network training. In: 2010 IEEE International Conference on Acoustics Speech and Signal Processing (ICASSP), pp. 4902–4905 (March 2010)
26. Schrauwen, B., Van Campenhout, J.: Improving spikeprop: enhancements to an error-backpropagation rule for spiking neural networks. In: Proceedings of the 15th ProRISC Workshop, vol. 11 (2004)
27. Stewart, R.D., Bair, W.: Spiking neural network simulation: numerical integration with the parker–sochacki method. Journal of Computational Neuroscience 27(1), 115–133 (2009)
28. Thomas, D.B., Luk, W.: Fpga accelerated simulation of biologically plausible spiking neural networks. In: 17th IEEE Symposium on Field Programmable Custom Computing Machines, FCCM 2009, pp. 45–52 (2009)
29. Yudanov, D., Shaaban, M., Melton, R., Reznik, L.: Gpu-based simulation of spiking neural networks with real-time performance & high accuracy. In: The 2010 International Joint Conference on Neural Networks (IJCNN), pp. 1–8 (February 2010)

An Iterative Feature Filter for Sensor Timeseries in Pervasive Computing Applications

Davide Bacciu

Dipartimento di Informatica, Università di Pisa, Italy
bacciu@di.unipi.it

Abstract. The paper discusses an efficient feature selection approach for multivariate timeseries of heterogeneous sensor data within a pervasive computing scenario. An iterative filtering procedure is devised to reduce information redundancy measured in terms of timeseries cross-correlation. The algorithm is capable of identifying non-redundant sensor sources in an unsupervised fashion even in presence of a large proportion of noisy features. A comparative experimental analysis on real-world data from pervasive computing applications is provided, showing that the algorithm addresses major limitations of unsupervised filters in literature when dealing with sensor timeseries.

1 Introduction

Pervasive computing puts forward a vision of an environment enriched by a distributed network of devices with heterogeneous sensing and computational capabilities, that are used to realize customized services supporting everyday activities. Pervasive computing systems deploy sensors that continuously collect data concerning the user and/or the environmental status. This data comes under the form of streams, i.e. timeseries, of sensor information with a considerably heterogeneous nature (e.g. temperature, presence, motion, etc.). This results in consistent amounts of information that need to be transferred and processed, typically in real time, to implement the system services, that are often realized by computational learning models (e.g. for predicting user activities based on sensed data) [1]. In this context, feature selection techniques for multivariate timeseries are fundamental, on one hand, to reduce the computational and communication overhead of transferring and processing such large amounts of sensor information. On the other hand, they serve to suppress redundant/irrelevant information which might negatively affect the predictive performance of the learning model.

We consider a pervasive learning system realized as part of the RUBICON project [2], that consists of a network of learning modules distributed on sensor motes characterized by limited computational and communication capabilities. Each of such device hosts a learning component implementing an Echo State Network (ESN) [3] model which is trained to perform real-time predictive tasks based on the data gathered by the sensors onboard the mote or received from another node through its radio interface. The optimization of the number of sensor streams feeding the learning modules is a key issue in such a resource constrained environment, requiring effective feature selection techniques

V. Mladenov et al. (Eds.): EANN 2014, CCIS 459, pp. 39–48, 2014.

for multivariate timeseries. Further, the RUBICON learning system allows to incrementally deploy new predictive tasks during system's operation, posing additional requirements on the feature selection model. The first is computational efficiency, as the selection process has to be performed in during system operation whenever a request for a new predictive task is posted. The second is the automatization of the feature selection process, as this has to be performed automatically by the learning system without any form of human/expert intervention (e.g to determine the number of selected features from a ranking).

Feature selection entails the identification of a subset of the original input sequences from a given dataset targeted at removing irrelevant and/or redundant information sources. In literature, the majority of the feature selection algorithms for multivariate timeseries take a wrapper approach, where the feature subset is selected to optimize the predictive and generalization abilities of a specific computational learning model [4]. Wrapper approaches are characterized by considerable computational requirements due to the burden of the multiple retraining of the underlying learning model and their results cannot be generalized to a different learning model. Corona (Correlation as Features) [5], for instance, is a wrapper method that transforms each multivariate timeseries into the corresponding correlation matrix, whose coefficients are fed to a support vector machine that is then used to apply the Recursive Feature Elimination method by [4]. Filter approaches, instead, use an external optimization criterion with respect to the learning model that will be using the selected data. Most of the filter techniques for timeseries data are tailored to classification tasks [6], as they select the streams that best separate multivariate samples from different classes. The Relief method, originally proposed for vectorial data, uses entropy as a measure of the ability of a feature to discriminate classes and has been extended to timeseries data [7]. The CleVer method [8] is one of the few unsupervised filter approaches for multivariate timeseries: it exploits the properties of the principal components common to all the timeseries to provide a ranking of the more informative features. Based on the assumption that there exists a common subspace across all multivariate data items, it first performs PCA on each univariate timeseries and then obtains the common principal components by bisecting the angles between their principal components. The CleVer method has found wide application, mainly due to its low computational requirements. However the number of selected features is not determined by the algorithm, rather it is selected by the user (as in k-means).

Previous works have noted how such sophisticated state-of-the-art feature selection techniques, which show excellent performances on multivariate timeseries benchmarks, do not provide significant results in context of open-ended discovery in real-world scenarios comprising a sensor-rich environment [9]. Motivated by this, we propose a simple, yet effective, feature selection technique based on a cross-correlation analysis of multivariate sensor timeseries, that is specifically tailored to the identification and removal of redundant sensor streams in an autonomous fashion. The proposed approach is based on an iterative filter heuristics that incrementally removes/selects timeseries based on redundancy information.

The algorithm is characterized by reduced computational requirements and by the ability to cope with the heterogeneous information sources that characterize a pervasive sensor system. Further, the feature selection process does not require expert intervention to determine the number of selected features and can therefore be fully automatized in the distributed learning system. The performance of the proposed feature selection approach is assessed on real-world data from indoor pervasive computing scenarios.

2   Iterative Sensor Timeseries Selection

We introduce the *Incremental Cross-correlation Filter* (ICF) algorithm for feature subset selection on multivariate timeseries (MTS) of sensor data. The ICF algorithm targets the reduction of the feature redundancy measured in terms of their pairwise cross-correlation. Let us define an univariate timeseries x^n as a the sequence of observations

$$x^n(1), \ldots, x^n(t), \ldots, x^n(T^n),$$

where $x^n(t)$ is the observation at time t of the n-th sample timeseries, and T^n is the sequence length. A D-dimensional MTS $\mathbf{x}^n$ is a collection of D univariate timeseries $x_i^n(1), \ldots, x_i^n(T^n)$ $(i = 1, \ldots, D)$, such that $x_i^n(t)$ is the observation at time t of the i-th component of the n-th sample MTS. In the following, we use the terms feature and variable to refer to a component of the MTS: each feature i is then associated to a set of univariate timeseries, one for each sample n.

The cross-correlation of two discrete timeseries x^1 and x^2 is a measure of their similarity as a function of a time lag (offset) τ, calculated through the sliding dot product

$$\phi_{x^1 x^2}(\tau) = \sum_{t=\max\{0,\tau\}}^{\min\{(T^1-1+\tau),(T^2-1)\}} x^1(t-\tau) \cdot x^2(t), \tag{1}$$

where $\tau \in [-(T^1-1), \ldots, 0, \ldots, (T^2-1)]$ and T^1, T^2 are the timeseries lengths. Intuitively, the lag where the maximum of the cross-correlation is computed provides information about the displacement between the first timeseries and the second.

The cross-correlation in (1) tends to return large numbers for signals whose amplitude is larger: this would prevent from comparing timeseries from different sensor modalities due to the considerably different scales of the sensor readings. To this end, we introduce the *normalized cross-correlation*

$$\overline{\phi}_{x^1 x^2}(\tau) = \frac{\phi_{x^1 x^2}(\tau)}{\phi_{x^1 x^1}(0) \cdot \phi_{x^2 x^2}(0)}, \tag{2}$$

where $\phi_{xx}(0)$ denotes the zero-lag autocorrelation, i.e. the correlation of a timeseries x with itself. The normalized function $\overline{\phi}_{x^1 x^2}(\tau)$ takes values in $[-1, +1]$, where a value of $\overline{\phi}_{x^1 x^2}(\tau) = 1$ denotes that the two timeseries have the exact same shape if aligned at time τ. Similarly, a value of $\overline{\phi}_{x^1 x^2}(\tau) = -1$ indicates that the timeseries have the same shape but opposite signs, while $\overline{\phi}_{x^1 x^2}(\tau) = 0$ denotes complete signal uncorrelation. From our point of view, both negative and positive extremes denote a certain redundancy in the information captured by the two timeseries. Therefore, the correlation value at the point in time where the signals of the two timeseries are best aligned is

$$\overline{\phi}^{*}_{x^1 x^2} = \max_{\tau} |\overline{\phi}_{x^1 x^2}(\tau)|. \tag{3}$$

The ICF algorithm implements a forward selection-elimination procedure that filters out redundant features, where redundancy is measured by the normalized cross-correlation in (2). ICF is based on the iterative application of a set of four selection/elimination rules, backed-up by the following intuitions

- A variable that is not correlated with any of the other features, should be selected.
- A variable that is correlated with all the variables that have already been selected is a good candidate for elimination.
- If the selection/elimination rules result in a working set of mutually correlated variables, act conservatively and maintain all those features that are less correlated with the selected ones.

The ICF algorithm is articulated in two phases: the former measures feature redundancy, while the latter iteratively applies the selection rules until all features are assigned to either the selected or the deleted status.

The first ICF phase builds a matrix of feature redundancy $R \in \{0,1\}^{D \times D}$, such that $R_{ij} = 1$ if features i and j are pairwise redundant and $R_{ij} = 0$ otherwise. Given a MTS dataset, the redundancy matrix is computed as follows

1. For each sample $\mathbf{x}^n$, use (3) to compute the maximum cross-correlation between all univariate sequences x_i^n, x_j^n in $\mathbf{x}^n$. If $\overline{\phi}^{*}_{x_i^n x_j^n} \approx 1$ for the pair i, j, assume the features i and j are correlated on the n-th sample.
2. Compute the percentage of samples in which each pair i, j is correlated.
3. Set $R_{ij} = 1$, if the corresponding feature pair i, j, with $i \neq j$, is correlated on more than $\theta_P \%$ samples.
4. Set the diagonal of R to zero, i.e. $R_{ii} = 0$ for all features i, to discount trivial correlations.

The redundancy matrix provides a unified picture of which variables are mutually correlated on a sufficiently large share of input samples. Experimentally, we have determined that a value of $\theta_P \% = 20\%$ is already sufficient to detect redundancies in a variety of experimental scenarios (nevertheless the value can also be determined on a per-task basis though cross-validation). Note that numerical issues discourage from using the exact $\overline{\phi}^{*}_{x_i^n x_j^n} = 1$ match in item 2 above: here, we suggest to consider a pair i, j to be correlated if $\overline{\phi}^{*}_{x_i^n x_j^n} > 0.99$.

The second phase applies the feature selection/elimination rules exploiting the information in the redundancy matrix R. It defines a set of unassigned features $\mathcal{F}$, that initially contains all the variables. The rules are applied iteratively to $\mathcal{F}$ following a priority order, until all features are assigned to either the set of selected variables $\mathcal{SF}$ or to the set of the deleted ones $\mathcal{DF}$. The details of the ICF rules and their priority pattern are described by the following procedure

1. RULE 0 - If a row R_i. is completely uncorrelated with the others in R (i.e. R_i. contains only zeros)
   (a) Add i to the selected subset: $\mathcal{SF} = \mathcal{SF} \cup \{i\}$;
   (b) Remove i from $\mathcal{F}$ and remove the corresponding entries in R;
   (c) If an uncorrelated feature j is generated as result of the previous step, move j from $\mathcal{F}$ to $\mathcal{DF}$ and remove the corresponding entries in R.
2. RULE 1 - If a row R_i. is correlated with all the others and R is not a matrix of all ones:
   (a) Add i to the deleted subset: $\mathcal{DF} = \mathcal{DF} \cup \{i\}$;
   (b) Remove i from $\mathcal{F}$ and remove the corresponding entries in R.
3. RULE 2 - If all features in $\mathcal{F}$ are mutually correlated with each other, i.e. R contains only ones,
   (a) Select the feature i that is less correlated with those currently in $\mathcal{SF}$;
   (b) Add i to the selected subset: $\mathcal{SF} = \mathcal{SF} \cup \{i\}$;
   (c) Remove i from $\mathcal{F}$;
   (d) Move the remaining features $\mathcal{F}$ to the deleted subset ($\mathcal{DF} = \mathcal{DF} \cup \mathcal{F}$) and terminate.
4. RULE 3 - If neither RULE 1 nor RULE 2 apply,
   (a) Extract i that is correlated with the minimum number of features in $\mathcal{F}$;
   (b) Define $\mathcal{S}(i) \subset \mathcal{F}$ as the subset of features correlated with i and select $j \in \mathcal{S}(i)$ as the maximally correlated feature with those currently in $\mathcal{SF}$;
   (c) Add i to $\mathcal{SF}$ and j to $\mathcal{DF}$;
   (d) Remove i, j from $\mathcal{F}$ and remove the corresponding entries in R.

The rationale of step 1(c) is that a feature j encoding the same information of already selected variables has to be deleted to avoid to be selected by future steps (otherwise RULE 0 is likely to be applied to j at the following iteration). Note that, in step 3(a), we determine the feature i that is minimally correlated with those in $\mathcal{SF}$ by measuring the pairwise cross-correlation between i and all $j \in \mathcal{SF}$, averaged across all samples, i.e.

$$\sum_{n=1}^{N} \frac{\overline{\phi}^{*}_{x_i^n x_j^n}}{N},$$

where N is the number of multivariate timeseries in the dataset. A similar approach is applied to steps 4(a) and 4(b).

The computational complexity of the ICF algorithm is, in general, dominated by the computation of the redundancy matrix in the first phase which strongly depends on the cost of computing the pairwise cross-correlation on the sample MTS. The asymptotic complexity of redundancy matrix computation is

$$O(N \cdot (D^2 \cdot T^{max})),$$

where N is the dataset length and the second term results from the computation of pairwise cross-correlations between D univariate timeseries with a maximum length T^{max}. The second phase of the algorithm is very efficient, i.e. linear in the number of features D, as the forward selection scheme processes each variable, at most, once with constant time operations. Therefore, the final complexity of the ICF algorithm is $O(N \cdot (D^2 \cdot T^{max}) + D)$.

3 Experimental Evaluation

The experimental evaluation is intended to assess the capability of the ICF algorithm in detecting and removing redundant MTS features in an indoor pervasive computing scenario. In particular, we compare the performance of ICF with respect to the CleVer method, a state of the art unsupervised feature filter for timeseries, as a function of the number of irrelevant features in the original MTS[1]. To this end, we have employed real-world data collected in two tasks involving the prediction of robot navigation preferences in a sensorized home-environment. The idea underlying these tasks is to learn to predict which navigation system is best to use to perform a certain trajectory based on environment characteristics and on user preferences. The preference weight to be learned is a value in [0, 1], where 1 is interpreted as maximum confidence on the navigation system and 0 denotes the lowest preference (i.e. the navigation system should not be used). The resulting computational learning task is, basically, a regression problem between the multivariate input timeseries and the corresponding univariate sequence of preference weights. For the purpose of feature selection evaluation, we only consider the input information (i.e. the sensor readings and trajectory information) but we discard the target data (i.e. the preference weight) as we are interested in assessing unsupervised selection methods.

The experimental scenario has been designed and put into operation in the Ängen senior residence facilities in Örebro Universitet. The scenario, depicted in Fig. 1, comprises a real-world flat sensorized by an RFID floor, a mobile robot with range-finder localization and a Wireless Sensor Network (WSN) with six mote-class devices, where the term M_i is used to denote the i-th mote. Each device is equipped with light (L), temperature (T), humidity (H) and passive infrared (P) presence sensors. The input information sources include all sensors from the six motes, plus robot trajectory information under the form of its (x, y) position and orientation θ, for a total of 24 features.

As shown in Fig. 1, the experimental assessment involves two tasks. The Entrance task is intended to predict a weight evaluating the performance of the localization system on two different trajectory types, represented as dashed and

---

[1] Matlab code for ICF and CleVer available at www.di.unipi.it/~bacciu/icf

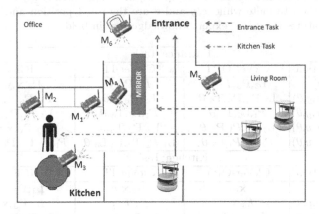

Fig. 1. Experimental scenario for the Entrance and Kitchen tasks in the Ängen facilities: M_i denotes the i-th WSN mote (Telosb platform running TinyOS)

continuous lines in Fig. 1. Performance on the dashed trajectory is expected to be low due to the effect of mirror disturbances which, conversely, should not affect trajectories on the continuous line. For the purpose of feature selection, the only relevant information is robot position and orientation (x, y and θ) as well as the P sensors onboard motes M_3 and M_6 (referred to as P_3 and P_6, respectively), that are the only presence sensors triggered by robot motion. The remainder of the sensors collect data that is poorly informative as it does not undergo significant changes across the timespan of data collection. The Kitchen task concerns a single trajectory type (dash-dotted arrows in Fig. 1) heading to the kitchen, where a user might be present or not. Since the robot range-finder localization is based on camera, the user is willing to switch it off every time he/she is in the room with the robot (the corresponding example trajectories are then marked with minimal preference, i.e. 0). The target of this task is to learn this user preference based on robot trajectory information and on the user presence pattern captured by the P sensors. The relevant information for this task is robot x-position (orientation and y coordinates do not change for this trajectory type) as well as the P sensors onboard motes M_1 to M_5 (i.e. P_1 to P_5), that are the only presence sensors that are triggered by robot or human motion. A total of 87 and 104 sequences have been collected for the two tasks sampling at 2Hz with an average length of 127 and 197 elements.

Table 1 shows the information sources selected by the CleVer and ICF algorithms for varying input configurations comprising different number of features: the selected relevant features are highlighted in bold. Since the CleVer algorithm requires the user to determine the number of selected features, we provide two set of results: one (CleVer-OPT) using the (known) optimal number of relevant features; the second (CleVer-ID) using the number of features found by ICF on the same configuration. Figure 2(a) provides a quantitative evaluation of the performance on the two tasks in terms of precision and recall of the selected

Table 1. Feature selection result for varying input configurations: M_i denotes all the transducers in the i-th mote while x, y and θ are the robot position and orientation. The relevant features (based on expert knowledge) are in bold.

Entrance Task			
Configuration	CleVer-OPT	CleVer-IT	ICF
(M_3,x,y,θ)	$L_3,\mathbf{P_3},T_3,\mathbf{y}$	$L_3,\mathbf{P_3},\mathbf{y}$	$\mathbf{P_3},\mathbf{x},\mathbf{y}$
(M_3,M_6,x,y,θ)	$L_3,\mathbf{P_3},\mathbf{P_6},T_6,\boldsymbol{\theta}$	$L_3,\mathbf{P_6},\mathbf{x},\boldsymbol{\theta}$	$\mathbf{P_6},\mathbf{P_3},\mathbf{x},\mathbf{y}$
$(M_4\text{-}M_6,x,y,\theta)$	$\mathbf{P_4},L_6,T_6,\boldsymbol{\theta}$	$\mathbf{P_4},L_6,T_6,\boldsymbol{\theta}$	$L_4,\mathbf{P_4},\mathbf{P_6},\mathbf{y}$
$(M_3\text{-}M_6,x,y,\theta)$	$L_3,T_3,\mathbf{P_4},\mathbf{P_6},\boldsymbol{\theta}$	$L_3,\mathbf{P_3},T_5,\boldsymbol{\theta}$	$\mathbf{P_3},\mathbf{P_6},\mathbf{x},\mathbf{y}$
$(M_1\text{-}M_6,x,y,\theta)$	$\mathbf{P_2},L_3,\mathbf{P_4},L_5,\boldsymbol{\theta}$	$L_1,\mathbf{P_2},T_2,L_3,L_6,\boldsymbol{\theta}$	$\mathbf{P_1},\mathbf{P_2},\mathbf{P_3},\mathbf{P_6},\mathbf{x},\mathbf{y}$
Kitchen Task			
Configuration	CleVer-OPT	CleVer-IT	ICF
(M_3,x,y,θ)	$\mathbf{x},\mathbf{y}$	$L_3,\mathbf{x},\mathbf{y}$	$L_3,\mathbf{P_3},\mathbf{x}$
(M_1,M_3,x,y,θ)	$L_3,\mathbf{x},\mathbf{y}$	$L_3,T_3,\mathbf{y}$	$L_1,\mathbf{P_3},\mathbf{x}$
$(M_1\text{-}M_3,x,y,\theta)$	$L_1,L_2,H_2,\mathbf{x}$	$L_1,L_2,H_2,\mathbf{x}$	$\mathbf{P_1},\mathbf{P_2},\mathbf{P_3},\mathbf{x}$
$(M_1\text{-}M_5,x,y,\theta)$	$L_1,\mathbf{P_2},L_3,T_3,L_4,H_5$	L_1,L_2,H_2,L_3,T_3,L_4	$\mathbf{P_1},\mathbf{P_2},\mathbf{P_3},\mathbf{P_4},\mathbf{P_5},\mathbf{x}$
$(M_1\text{-}M_6,x,y,\theta)$	$T_1,L_2,H_2,L_3,\mathbf{P_4},\mathbf{P_6}$	$T_1,L_2,\mathbf{P_2},H_2,L_3,\mathbf{P_4}$	$\mathbf{P_1},\mathbf{P_2},\mathbf{P_3},\mathbf{P_4},\mathbf{P_5},\mathbf{x}$

features: e.g. precision is the proportion of correctly selected features (true positives) with respect to the total number of selected features (true and false positives); similarly for recall. The results show that ICF is capable of consistently reducing the number of input features by maintaining the majority (if not all) of the relevant features even when a large number of uninformative features is included. Conversely the performance of both the CleVer methods deteriorates consistently as the proportion of redundant features increases. This behavior is clear in the precision-recall curves in Fig.2(a): for ICF, the proportion of false positives (FP) and negatives (FN) does not grow with the size of the search space (and the number of potentially irrelevant features); whereas, both Clever methods experience a marked performance deterioration due to an increase in both the FP, inducing a reduction in the precision values, as well as in the FN, which lowers the methods recall. Additionally, the results on last four configurations of the Entrance task in Table 1 show that the CleVer algorithm yields to different features subsets for different repetitions (note that the number of selected features in CleVer-OPT and CleVer-ID is the same for these configurations). This is the result of the well-known sensitivity to initialization of the k-means algorithm used by CleVer. ICF, on the other hand, has a stable behavior yielding to the selection of the same feature subset for multiple algorithm repetitions.

Representation entropy provides a means for evaluating the effectiveness of algorithms in terms of amount of redundancy present in the selected feature subsets. Let X be the $K \times K$ covariance matrix of the K selected features and λ_i the eigenvalue associated to the i-th feature, define $\bar{\lambda}_i = \lambda_i / \sum_{j=1}^{K} \lambda_j$. Then, the representation entropy is $E_R = -\sum_{i=1}^{K} \bar{\lambda}_i \log \bar{\lambda}_i$, such that E_R attains its minimum when all the information is concentrated along a single feature, making the rest redundant. Figure 2(b) shows the E_R value on the Kitchen task as a function of the input configuration: ICF confirms its ability to identify and filter-out redundant information yielding to the best performance when the proportion

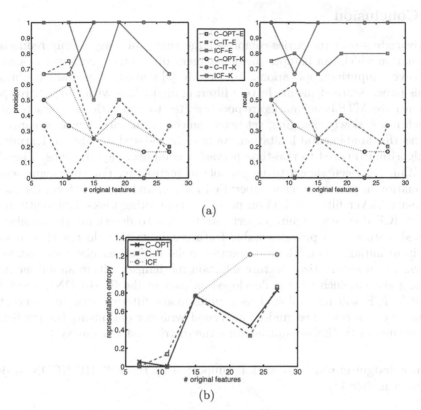

(a)

(b)

Fig. 2. Quantitative feature selection performance: 2(a) show the precision and recall of the selected features as a function of the original input space size for the Entrance (square placeholders) and Kitchen (diamond markers) tasks. Clever algorithms are identified as C-OPT and C-IT: note that the C-OPT-E recall curve is completely overlapping with that by C-ID-E. Fig. 2(b) shows the representation entropy of the three methods on the Kitchen task.

of noisy features is high. To understand the impact of noisy and redundant features on the final learning task, consider that an ESN with 500 reservoir neurons trained (i.e. resulting from model selection on a validation set) on the Kitchen task using all the WSN inputs (i.e. without feature filtering) achieves a Mean Absolute Error (MAE) on the test set that is ≈ 0.5. The corresponding ESN using only the ICF filtered inputs reported in the last row of Table 1 achieves a test MAE of ≈ 0.015, reducing the original unfiltered error by 97%.

ICF effectiveness is not obtained at the cost of its computationally efficiency: e.g. the average time required to complete feature selection is 2153msec for the fifth configuration of the Entrance task (Java code running in an Eclipse box on an Intel I5 Quad-core at 2.7 GHz CPU equipped with 4GBytes of RAM). The majority of the running time is spent on redundancy mask computation, while feature filtering effort is negligible, i.e. 1msec.

4   Conclusion

Multivariate sensor timeseries comprise large shares of noisy, highly redundant information which can hamper the deployment of effective predictive models in pervasive computing applications. As noted in [9] and experimentally confirmed in this paper, state-of-the-art feature filtering algorithms with competitive performances on MTS benchmarks are poorly suited to deal with the characteristics of such noisy, slowly changing, yet heterogeneous in nature, sensor streams. To address this fundamental limitation, we have introduced an efficient feature filter algorithm tailored to real-time pervasive computing applications. The ICF algorithm has been shown to be capable of identifying non-redundant sensor information in a completely unsupervised fashion and to outperform the state-of-the-art CleVer filter method on pervasive computing tasks. Differently from CleVer, ICF does not require expert intervention to determine the number of selected features and provides stable feature subsets that do not change with algorithm initialization. These properties make ICF an excellent candidate to implement an automatized feature selection mechanism within an autonomous learning system, such as that developed as part of the RUBICON project [2]. As such, ICF will be exploited as a preliminary filtering step to reduce the complexity of a relevance-guided supervised wrapper optimizing the predictive performance of the ESNs implementing the distributed learning system.

Acknowledgements. This work is supported by the FP7 RUBICON project (contract n. 269914).

References

1. Ye, J., Dobson, S., McKeever, S.: Review: Situation identification techniques in pervasive computing: A review. Pervasive Mob. Comput. 8(1), 36–66 (2012)
2. Bacciu, D., Barsocchi, P., Chessa, S., Gallicchio, C., Micheli, A.: An experimental characterization of reservoir computing in ambient assisted living applications. Neural Computing and Applications, 1–14 (2013)
3. Jaeger, H., Haas, H.: Harnessing nonlinearity: Predicting chaotic systems and saving energy in wireless communication. Science 304(5667), 78–80 (2004)
4. Guyon, I., Weston, J., Barnhill, S., Vapnik, V.: Gene selection for cancer classification using support vector machines. Mach. Learn. 46(1-3), 389–422 (2002)
5. Yang, K., Yoon, H., Shahabi, C.: A supervised feature subset selection technique for multivariate time series. In: Proc. of FSDM 2005, pp. 92–101 (2005)
6. Han, M., Liu, X.: Feature selection techniques with class separability for multivariate time series. Neurocomput. 110, 29–34 (2013)
7. García-Pajares, R., Benítez, J.M., Sainz-Palmero, G.: Frasel: a consensus of feature ranking methods for time series modelling. Soft Computing 17(8), 1489–1510 (2013)
8. Yoon, H., Yang, K., Shahabi, C.: Feature subset selection and feature ranking for multivariate time series. IEEE Trans. Knowl. Data Eng. 17(9), 1186–1198 (2005)
9. Cheema, S., Henne, T., Koeckemann, U., Prassler, E.: Applicability of feature selection on multivariate time series data for robotic discovery. In: Proc. of ICACTE 2010., vol. 2, pp. 592–597 (2010)

Exploiting Evolution on UAV Control Rules for Spraying Pesticides on Crop Fields

Bruno S. Faiçal[1], Gustavo Pessin[2], Geraldo P.R. Filho[1], Gustavo Furquim[1], André C.P.L.F. de Carvalho[1], and Jó Ueyama[1]

[1] Institute of Mathematics and Computer Science (ICMC)
University of São Paulo (USP) - São Carlos, SP, Brazil
{bsfaical,geralop,furquim,andre,joueyama}@icmc.usp.br
[2] Vale Institute of Technology
Belém, PA, Brazil
gustavo.pessin@itv.org

Abstract. The application of chemicals in agricultural areas is of crucial importance for crop production. The use of aircrafts is becoming increasingly common in carrying out this task mainly because of their speed and effectiveness. Nonetheless, some factors may reduce the yield, or even cause damage, like areas not covered in the spraying process or overlapped spraying areas. Weather conditions add further complexity to the problem. Sets of control rules, to be employed in an autonomous Unmanned Aerial Vehicles (UAV), are very hard to develop and harder to fine-tune to each environment characteristics. Hence, a fine-tuning phase must involves the parameters of the algorithm, due to the mechanical characteristics of each UAV and also must take into account the type of crop being handled and the type of pesticide to be used. In this paper we present an evolutionary algorithm to fine-tune sets of control rules, to be employed in a simulated autonomous UAV. We describe the proposed architecture and investigations about changing in the evolutionary parameters. The results show that the proposed evolutionary method can fine-tune the parameters of the UAV control rules to support environment and weather changes in the simulated environment, encouraging the deployment of the system with real hardware.

1 Introduction

Chemical defensives, also known as pesticides, are commonly applied in agricultural areas to increase productivity. However, these products can cause serious health problems for workers who have direct or indirect contact with them. There are various diseases that can result from the interaction with these chemicals, like cancers, complications in the respiratory system and neurological diseases [15]. It is estimated that about 2.5 million tons of pesticides are applied worldwide each year and that this amount has been growing [12]. Much of the pesticide is lost during the spraying process due to the type of technology employed. Nevertheless, only a small part of the pesticide reaches the target crop field while the rest of it drifts away [10]. Evidences of pesticide drifts are commonly found

V. Mladenov et al. (Eds.): EANN 2014, CCIS 459, pp. 49–58, 2014.
© Springer International Publishing Switzerland 2014

between 48 m and 800 m from the target crop field. Other problems are crop areas not covered in the spraying process and overlapped spraying areas.

The use of UAVs to carry out the task of spraying pesticides can be beneficial to many reasons, including (i) to reduce human contact with the chemicals, which helps to preserve human health; and (ii) to improve the performance of the spraying operation, avoiding the presence of chemicals outside designed areas, which helps to preserve neighborhood fields, that can be other crops, preserved nature areas or water sources. Sets of control rules, to be employed in an autonomous UAV, are very hard to develop and harder to fine-tune to each environment characteristics. Thus, a fine-tuning phase must involves the parameters of the algorithm, due to the mechanical characteristics of each UAV and also must take into account the type of crop being handled and the type of pesticide to be used. In this paper we present a evolutionary algorithm to fine-tune sets of control rules, to be employed in an simulated autonomous UAV. We describe the proposed architecture and investigations about changing in the evolutionary parameters.

The proposed architecture employs an UAV, which has a system of coupled spray, and it is able to communicate with the Wireless Sensor Network (WSN), which is organized in a matrix-like disposition on the crop field. This WSN aims to send feedback on the weather conditions and how spraying actually are falling in the target crop field. Based on the information received, the UAV appropriately applies a policy to correct its route. Hence, the main contributions of this research are as follows: (i) investigate an evolutionary methodology capable of minimize human contact with pesticides, (ii) evaluate an evolutionary approach able to minimize the error in spraying pesticides in areas of growing vegetables and fruits, (iii) investigate techniques able to maximize quality in agricultural production, and (iv) contribute to increase the autonomy of the architecture proposed by [5], in which the policy parameters were set empirically and applied independent of weather conditions.

This paper is organized in 5 sections. Section 2 presents other studies related to this paper. The proposed methodology is described in Section 3. Results from investigations are presented in Section 4. In Section 5 we present a discussion upon the results; this section also presents the conclusions and describes some future work.

2   Related Work

There are several works that employs UAVs as agents in agriculture and WSN as monitors of the environment, occasionally integrating both [2,7,16]. For example, Huang and collaborators [6] propose a system for spraying pesticide coupled to an UAV capable of carrying as much as 22.7 kg. The UAV model used was a SR200 manufactured by Rotomotion company. The spray system consists of four major components: (i) a metal tube with nozzles, (ii) a tank to store the pesticide, (iii) a pump to move the liquid and (iv) a mechanism for controlling the activation of spray. The spraying system can carry up to 5 kg of pesticide, which was needed to spray 0.14 km^2; and it provides a flight time of around 90 minutes. The main objective of that work was to validate the proposed system and evaluate different spray nozzles. However, the weather conditions were not taken into account. Additionally, a discussion of the evolutionary methodology able to optimize control of this activity is not presented.

Valente and collaborators [13] show a system based on WSNs and UAVs to monitor crop fields of vines. The WSN collects information from soil, climate and the condition of vines and presents this data to the farmer. However, the vine crop groups may be hundreds of meters distant from each other. Because of barriers (eg. rivers and roads) that may occur between crop fields, the usage of cables to connect networks implies in a prohibitive cost. Although the use of more powerful radios in sensor nodes enables communication between WSN, this will result in higher energy consumption implying in the reduction of battery lifetime. Thus, the solution used to overcome such limitations was employ a UAV able to fly over crop fields and collecting the information from each WSN, bringing data back to a processing center.

Faiçal and collaborators [5] proposed and evaluated an architecture formed by UAV and WSN to spray pesticides in crop fields. It is known that the weather conditions in the area of cultivation, such as wind speed and direction, can cause error in the spraying process. The study showed that the proposed architecture allows to minimize error and increase control of this activity. However, the work used a simplistic approach to correct the route of the UAV. The parameters set for the correction of the route are similarly applied in different weather conditions, which can harm the performance of this architecture. As previously mentioned, the objective of this paper is to evaluate and propose an evolutionary methodology to optimize and define the best weather parameter that influences the intensity correction of the UAV route.

3 Methodology

Fig. 1 synthesizes the context of this work. It can be seen that the spraying is carried out using UAVs, which have equipment for pesticide spraying, and a WSN distributed in matrix disposition in the crop field. The WSN is represented only in a target crop field delimited with two dashed lines (from the upper left to the lower right corner) to simplify the visualization. The two arrows indicate the direction of the wind at a specific location. The UAV maintains communication with WSN about the weather conditions (wind speed and direction) in its current position and also about the concentration of the pesticides identified by surrounding sensors. When an imbalance in the pesticide concentration is detected (e.g. the sensor on the left side identified a higher concentration than the one positioned on the right side), possibly caused by winds, the UAV uses its policy to change the position so that the pesticide is applied at a concentration balanced across the width of the target crop field. In addition, constraints prevent the pesticide to be sprayed out of the bounds of the target crop field, which may cause an overlap of the area that was subjected to the defensive chemicals. The adjustment of the route is represented by small arrows between the images

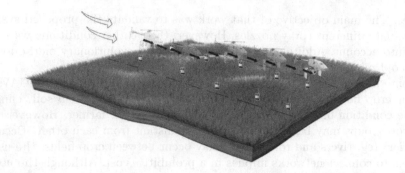

Fig. 1. Spraying using the architecture proposed by [5]. This architecture is formed by a UAV (spray) and WSN (sensing and feedback). If spraying is unbalanced between the sensors (distributed in matrix form on crop field), the UAV can correct your route using a policy with parameter settings defined before starting the activity.

of the UAV in Fig. 1. To adjust the route, the policy has the *routeChangingFactor* parameter, which operates pondering the intensity of alteration (sudden or soft). This parameter is set empirically before the activity and will be constant for all weather conditions.

In this work we extend the architecture proposed by [5], adding an evolutionary module able to optimize the parameter *routeChangingFactor*. Furthermore, the UAV will query the WSN about weather conditions of a target crop field. With this information, the UAV simulates computationally the result of spraying using different possible configurations. These simulations take into account the weather conditions informed by WSN and settings of UAV. The source code contains all instructions necessary to simulate the behavior and communication between the UAV and WSN. It also contains a dispersion model to represent the movement of sprayed particles along the crop field. These simulations use a Genetic Algorithm (GA) which evaluates the results and evolves to find a near-optimal *routeChangingFactor* to be used. This optimization is carried out for each target crop field until the whole desired area is sprayed. It is worth mentioning that the optimization is carried out in parallel to the spraying of pesticides and the *routeChangingFactor* is changed only when the GA finalize and the UAV enters the analyzed crop field.

To investigate the evolution in control we considered a rectangular field 1100 m long and 150 m wide. Moreover, a target crop field was considered to cover a rectangle measuring 1000 m by 50 m. The WSN consists of 20 sensor nodes arranged in matrix form throughout the target crop field. The UAV flies 20 feet high at a constant speed of 15 m/s, communicating with the WSN every 10 seconds.

The fitness function is the sum of pesticide gathered by the WSN outside the boundaries, greater the number means that greater amount of pesticide was placed outside the boundaries; hence, this fitness should be minimized. Current genome has a single real value that represent the *routeChangingFactor*; it is detailed in the next section. We treated the genome as a real value because it could be directed applied to the simulated UAV as an value to the rotors.

3.1   Deployment of the Evolutionary Module

Projects on WSN and UAVs are commonly validated in two ways: (i) testbeds and/or (ii) simulation. The testbed is a smaller version of the project built to conduct experiments. On the other hand, simulation is the act of using computers in formalization, as mathematical expressions or specifications, to mimic a real-world process. The scientific community has used the simulation method to validate WSN environments before real deployment [3,9]. Results from simulation are considered satisfactory in comparison to the results obtained from testbeds [1,8]. Thus, simulation results may be used to justify changes in order to minimize the negative impact in a real environment.

The same platform from [5] was employed to run the simulations. The OM-NeT++[1] is a discrete event simulator based on C++ to model communication networks, multiprocessors and other parallel and distributed systems [14]. The OMNeT++ has a wide scope so it can be used to simulate various types of networks. The GA is configured to use a crossing value of 90% in the population and apply a mutation of 10%, besides employing the technique of elitism (where the best individual is kept for the next generation). Table 1 exemplifies the population used by the genetic algorithm, in which each individual is composed of a positive real value for the *routeChangingFactor* and its respective fitness which is calculated by adding all the particles of pesticide that are applied outside the target crop field. Therefore, a lower value of fitness indicates a better individual.

Table 1. Representation of the population used by the Genetic Algorithm

Individual	routeChangeFactor	Fitness
1st	2.136	12,032
2nd	2.532	12,169
3rd	1.465	20,032
4th	4.752	24,878
5th	3.846	22,987

In the experiments five population sizes and three maximum values of generations are evaluated. Each setting of experiments is represented by Ind**M**Ger**N**. Thus, **M** is the number of individuals of the population and **N** the value maximum of generations. As a stopping condition, we define the maximum amount of generations for each experiment, so after running all the pre-defined generations the GA is finished and the best individual of this generation is considered the *routeChangingFactor* more suitable for the weather conditions monitored by the WSN. Each configuration of the experiments were replicated 30 times in order to obtain a sample with high reliability to analyze its results.

The GA evolves the population according to their characteristics already described, and changes the configuration to be evaluated through the assignment

---

[1] OMNeT++ Network Simulation Framework, http://www.omnetpp.org

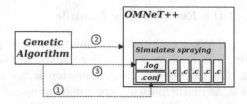

Fig. 2. Interaction between Genetic Algorithm and the simulator OMNeT++

of a new value to the *routeChangingFactor* variable (considering the individual to be tested) in the source code inside the simulator. Fig. 2 shows the interaction between the GA and the simulator. Initially, the GA alters the configuration file of module Simulates Spraying with the value of *routeChangingFactor* of individual to be evaluated (step 1). Subsequently, the GA run the module Simulates Spraying in OMNeT++ (step 2) and finally analyzes the file *log* of the executed plan (step 3). This file stores the result of spraying all over the field (1500 m x 150 m) and the amount of pesticide sprayed wrongly (outside the target crop field 1000 m by 50 m) is considered as fitness of the individual. When the GA has tested all individuals of a generation, it will produce a new generation of individuals until the maximum generation is reached. During this study, we tried to keep the GA simple and fast; this is important because all analysis need to be carried in short time, once the spray occurs at runtime.

4   Results

We employed the Genetic Algorithm as an evolutionary method to find the best *routeChangingFactor* to be used at a target crop field, considering the weather conditions identified by the WSN[2]. Fig. 3 shows three heat maps of sprays in the crop field. It can be observed that the target crop field is shown in this image, thus it is possible to identify where the pesticide was actually applied in or out of target field. The values 6.000 and 4.000 were defined empirically, whereas the value 2.140 was obtained by the proposed evolution module. Also is possible observe that the map of the spraying performed using the Evolution Module portrays a most appropriate correction of route considering weather conditions identified by WSN. This optimization provides a spray with lower error rate than the others (see Fig. 3), and provide a setting at runtime this parameter.

To evaluate the results, we performed a series of statistical analyzes. We started using the Shapiro Wilk method to verify the adequacy of normality and consequently to direct it to use parametric or non-parametric methods according to the results. We could observe that all values are less than 0.05, hence, all sets have the hypothesis of normality rejected considering a confidence level of 95%. Thus, we use non-parametric tests in the subsequent analyzes.

---

[2] Source code available in http://goo.gl/9S14TO

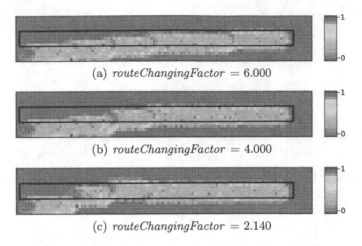

(a) *routeChangingFactor* = 6.000

(b) *routeChangingFactor* = 4.000

(c) *routeChangingFactor* = 2.140

Fig. 3. A heat map to represent the chemicals sprayed on the crop at the end of the simulation. The green colour represents no pesticide and red represents the most concentrated places. The thin black lines show the crop field that needs to have chemicals sprayed. (a) and (b) Evaluations with empirical values. (c) Evaluation with *routeChangingFactor* obtained by the genetic algorithm. We can see that when employing the *routeChangingFactor* obtained by the genetic algorithm we have the best adjusts in the UAV track, attempting to keep the chemicals within the boundary lane. It is worth to highlight that, as the simulation starts with wind, the UAV always starts the dispersion of the chemicals outside the boundary.

As implied in Fig 4(a), there appears to be an improvement in the obtained results (lower error) with the increase of individuals. The pairwise comparisons using Wilcoxon rank sum test shows that there is a significant difference in populations formed by 3, 5 and 10 Individuals but not for populations with 10, 15 and 20 Individuals. This may imply that there is need to further increase the number of generations. Figures 5(a) and 6(a) shows the results with 50 and 100 generations.

The pairwise comparisons using Wilcoxon rank sum test shows that for experiments using populations with 5, 10, 15 and 20 Individuals for 50 generations there is no significant difference. However, using populations with 5 and 10 individuals have lower accuracy populations when compared with the results of the populations with 15 and 20 Individuals. To the experiments with 5, 10, 15 and 20 Individuals and 100 generations no significant difference and their accuracies are similar. It should be noted that the settings used in the Genetic Algorithm provide results with high accuracies. Therefore, the settings that resulted in the best results are the populations formed with 15 and 20 Individuals to 50 Generations and 5, 10, 15 and 20 Individuals to 100 Generations.

From Fig. 4(b), 5(b) and 6(b), we can see that the average runtime time grows as the population and the amount of generations increase. Considering the settings that correspond to the best results (Ind15Ger50, Ind20Ger50, Ind5Ger100, Ind10Ger100, Ind15Ger100, Ind20Ger100), it is possible to note that the setting Ind5Ger100 has the lowest average runtime of 44.12 seconds. As described

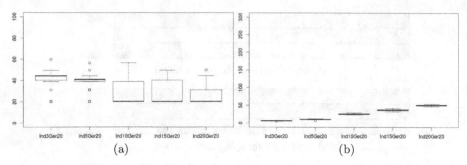

Fig. 4. Results of the GA employing 20 generations. (a) Fitness. (b) Time (in seconds).

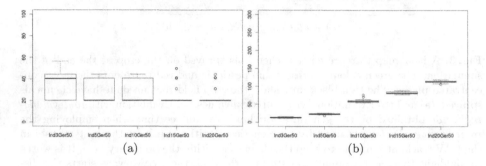

Fig. 5. Results of the GA employing 50 generations. (a) Fitness. (b) Time (in seconds).

above, the UAV flies at a speed of 15m/s in these experiments. Therefore, using the setting Ind5Ger100 to analyze the target crop field the UAV would fly over 661.907 meters. Thus, we can conclude that due to the length of target crop field measuring 1000 meters, this setting allows the later target crop field to be analyzed while the current target crop field is sprayed.

5   Discussion

We have described a methodology to evolve the parameter *routeChangingFactor*, which aims to adjust the UAV route and improve the spraying of pesticides on crop fields. The spraying operation is conducted employing an architecture based on a UAV and WSN. The UAV is the agent which spray the pesticide and the WSN is responsible for the monitoring of (i) weather conditions, (ii) points where the pesticide reached the crop field and (iii) feedback to the UAV. The initial methodology, although functional, have showed some limitations in correcting the route, since this parameter was defined empirically and remained the same for all activity. This limitations is corrected with the proposal described in this work.

Due to the fact that the adjustment of the route is performed using the *routeChangingFactor* parameter. It may be noted that in the experiments that

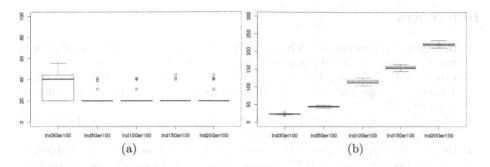

(a)   (b)

Fig. 6. Results of the GA employing 100 generations. (a) Fitness. (b) Time (in seconds).

we use 20 generations to find the best parameter, settings involving populations with 10, 15 and 20 individuals are not significantly different. However, only 50% of the results achieved is the best possible value. The same happens with the results obtained in experiments using populations with 5 and 10 individuals when evolved by 50 generations. Moreover, the results obtained using 15 and 20 individuals by 50 generations and also 5, 10, 15 and 20 individuals by 100 generations achieved the best possible value and showed greater stability in its results. Considering these results, the configuration Ind5Ger100, corresponding to a population of 5 individuals with 100 generations, has satisfactory behavior for the purpose this study. This configuration is able to achieve good results with high accuracy at relatively low average runtime. Other settings of the experiments not cited in this section are considered unsuitable for solving this problem because of its low accuracy.

It is worth mentioning that the error in no case is less than 20% because the methodology considers that the UAV starts spraying at a fixed point and the route adjustment occurs after some predefined time. Thus, this error occurs at the beginning of the crop field where spraying has the influence of the weather conditions. Therefore, a better understanding of the results is made from the following reading: starting spraying in position X, the best *routeChangingFactor* has value Y, which will result in an error of $Z\%$ in weather conditions informed by the WSN.

Lastly, it is important to remember that the developed methodology, which evolve the *routeChangingFactor*, had as main motivation the possibility of providing to UAV with a intelligent behavior, adjusting its route considering weather conditions. Thus, this becomes a dynamic policy for a naturally dynamic environment. The next stages of this project will be as follows: (i) developing the system using real hardware, addressing the reality gap in communications between the UAV and the WSN, the behaviour of the UAV and the sensor capabilities and (ii) investigating the use of other evolutionary techniques, like NSGA-II [4] and Differential Evolution [11]. As a final observation, since it is necessary to improve the simulation environment (which allows quicker and safer evaluations) other future work should seek to improve the current chemical dispersion module and the physical behaviour of the UAV.

References

1. Bergamini, L., Crociani, C.: Vitaletti: Simulation vs real testbeds: a validation of wsn simulators. Technical report n. 3, Sapienza Universita di Roma (2009)
2. Branco, K.R., Pelizzoni, J.M., Neris, L.O., Junior, O.T., Osorio, F.S., Wolf, D.F.: Tiriba - a new approach of uav based on model driven development and multiprocessors (2011)
3. Chen, H., Chang, K., Agate, C.S.: A dynamic path planning algorithm for uav tracking. In: SPIE Defense, Security, and Sensing (2009)
4. Deb, K., Pratap, A., Agarwal, S., Meyarivan, T.: A fast and elitist multiobjective genetic algorithm: Nsga-ii. IEEE Transactions on Evolutionary Computation 6(2), 182–197 (2002)
5. Faical, B.S., Costa, F.G., Pessin, G., Ueyama, J., Freitas, H., Colombo, A., Fini, P.H., Villas, L., Osorio, F.S., Vargas, P.A., Braun, T.: The use of unmanned aerial vehicles and wireless sensor networks for spraying pesticides. Journal of Systems Architecture 60(4), 393–404 (2014)
6. Huang, Y., Hoffmann, W.C., Lan, Y., Wu, W., Fritz, B.K.: Development of a spray system for an unmanned aerial vehicle platform. Applied Engineering in Agriculture 25(6), 803–809 (2009)
7. Li, B., Liu, R., Liu, S., Liu, Q., Liu, F., Zhou, G.: Monitoring vegetation coverage variation of winter wheat by low-altitude uav remote sensing system. Trans. of the Chinese Society of Agricultural Engineering 28(13), 160–165 (2012)
8. Malekzadeh, M., Ghani, A.A.A., Subramaniam, S., Desa, J.: Validating reliability of omnet++ in wireless networks dos attacks: Simulation vs. testbed. International Journal of Network Security 12(3), 193–201 (2011)
9. Ouyang, J., Zhuang, Y., Xue, Y., Wang, Z.: Uav relay transmission scheme and its performance analysis over asymmetric fading channels. Hangkong Xuebao/Acta Aeronautica et Astronautica Sinica 34(1), 130–140 (2013)
10. Pimentel, D.: Amounts of pesticides reaching target pests: environmental impacts and ethics. Journal of Agricultural and Environmental Ethics 8(1), 17–29 (1995)
11. Price, K.V., Storn, R.M., Lampinen, J.A.: Differential evolution a practical approach to global optimization (2005)
12. Tariq, M.I., Afzal, S., Hussain, I., Sultana, N.: Pesticides exposure in pakistan: A review. Environment International 33(8), 1107–1122 (2007)
13. Valente, J., Sanz, D., Barrientos, A., Cerro, J., Ribeiro, A., Rossi, C.: An air-ground wireless sensor network for crop monitoring. Sensors 11(6), 6088–6108 (2011)
14. Varga, A.: Omnet++. In: Wehrle, K., Günes, M., Gross, J. (eds.) Modeling and Tools for Network Simulation, pp. 35–59. Springer, Heidelberg (2010)
15. Weisenburger, D.D.: Human health effects of agrichemical use. Human Pathology 24(6), 571–576 (1993)
16. Xiang, H., Tian, L.: Development of a low-cost agricultural remote sensing system based on an autonomous unmanned aerial vehicle (uav). Biosystems Engineering 108(2), 174–190 (2011)

Fuzzy-Logic Decision Fusion for Nonintrusive Early Detection of Driver Fatigue or Drowsiness

Mario Malcangi

Department of Computer Science, Università degli Studi di Milano,
Via Comelico 39, 20135 Milano, Italy
malcangi@di.unimi.it

Abstract. Traffic accidents due to falling asleep at the wheel are a longstanding problem in many countries. This paper presents a novel solution based on fuzzy-logic decision fusion that prevents accidents by detecting driver fatigue or drowsiness early. The proposed method is based on analyzing and inferring about certain biological and behavioral measurements that enable detection of reduced alertness preceding driver-sleep onset. Because wakeful or sleep activity is reflected in several physiological conditions in human beings, such as cardiac, breathing, movement, and skin galvanic conductance, captured bioelectric signal features were extracted and fuzzy decision-fusion logic was tuned to make inferences about oncoming driver fatigue or drowsiness. The proposed method improves the performance by applying the fuzzy logic inference to fuse decisions from independent modules that infer about features measured on the sensed physiologic and/or behavioral information. The method reduces the complexity of the signal processing and of the pattern matching model. Tests have been executed on clinical and in field physiologic and behavioral data. A prototype based on a 32 bit microcontroller and a highly integrated analog front-end has been developed to support the in field tests.

Keywords: fuzzy logic, decision fusion, sleep onset, heart-rate variability, breathing rate, power-spectrum density, ANS.

1 Introduction

Driver fatigue or drowsiness is a cause of serious traffic accidents, so several methods have been investigated to find a practical solutions for early sleep-onset detection with the aim of achieving a higher level of safety in private and public transportation. According to some traffic authorities, crashes caused by drowsy drivers are extremely frequent, e.g. about 40,000 per year in the United States alone, [1] but such statistics underestimate the number because they do not include most crashes due to fatigue or where drowsiness was not evident, so this kind of risk really involves millions of drivers. If fatigue and drowsiness are detected early, accidents can be prevented by implementing countermeasures based on onboard safety devices, such as automatic cruise control, or simply on acoustic alarms if the system is not integrated into the onboard cruise electronics. To detect sleep onset early, the driver's state needs to be continuously monitored. Several investigations were carried out, following two main approaches, one based on monitoring driver behavior and the other on measuring driver physiology.

V. Mladenov et al. (Eds.): EANN 2014, CCIS 459, pp. 59–70, 2014.

1.1   Literature Review

Drowsiness is the first stage of non-rapid-eye-movement (NREM), so an approach to early sleep detection was based on image-processing techniques. Images of the driver's face are captured by a digital camera and processed to match behavior features such as eyelid movements, yawning, eye gaze, and head nodding [2]. Most studies related to behavior observation are based on image recognition that focuses on a small area of the face, especially the eyes, to measure blink rate. Degree of eyelid opening is considered a valid indicator of the tiredness level that leads toward the critical threshold of fatigue and drowsiness. Eyelid movements in normal conditions are large, quick, and constant, but, when fatigue or drowsiness is oncoming, such movements tends to become shorter, slower, and irregular. More information can be extracted from facial expression, when it changes during passage from alert status to drowsiness. Such an approach is noninvasive, and industrial implementations exist. However, it somewhat not reliable, due to the background noise, the driver movements, and the difficulty of focusing automatically on a limited part of the driver's face (near the eyes).

Drowsiness is controlled by the autonomous nervous system (ANS) acting on cardiac rhythm, breathing, and galvanic skin response, so an alternative approach to early detection of falling asleep was based on bioelectric signal processing. This approach is very effective, because it does not require that the driver be collaborative [3, 4, 5]. It needs to capture bioelectric signals from the heart, skin, and pulmonary apparatus by using a set of electrodes applied to the driver's body. Therefore, this approach is initially invasive, but this limitation can be overcome because cardiac rhythm and skin conductivity can be measured at the hands. A set of contact electrodes can be applied to the steering wheel, where the driver puts his hands. Breathing rate can be measured noninvasively if an appropriate sensor is built into the safety belt or the back of the driver's seat. Some behavior information, such as arm and steering wheel movements, can also be measured at the steering wheel.

Physiological and behavior measurements captured in the field without driver cooperation are extremely fuzzy. A smart system is needed to make inferences about sleep onset using physiology and behavior features extracted from signals captured by one or more hand-contact electrodes and some sensors, and then to fuse the data. A fuzzy-logic-based inference system can be very effective, if several physiological and behavioral features concur in the decision.

1.2   Novelty and Advantages of the Proposed Method

A practical system for detecting sleep onset early can be based on ECG measurements alone, because there is enough information in the ECG signal directly related to sleep-wake control [6]. Nevertheless, a multimodal approach would improve early drowsiness detection and might also enable evaluation of fatigue level rising toward a critical threshold.

Sleep is a physiological state characterized by variation in ANS activity, which is reflected in heart-rate variability (HRV). The power spectral density (PSD) of heart rate varies with the change from wakefulness to sleep [7,8]. The low-to-high frequency ratio is a valid indicator of such change, because it reflects the balancing action of the ANS's sympathetic and parasympathetic nervous-system branches. When sympathetic nervous-system activity increases, parasympathetic nervous-system activity diminishes, thus causing cardiac rhythm to accelerate (shorter beat intervals). Cardiac-rhythm deceleration is caused by low sympathetic nervous-system activity and increased parasympathetic nervous-system activity, producing heart-rhythm deceleration (longer beat intervals).

PSD analysis of beat-to-beat HRV (Fig. 1) is useful to understand when sleep is setting in. Sleep and wakefulness are directly related to the ANS [9]. If we consider low-frequency versus high-frequency balance in a person's PSD, it is possible to predict the onset of sleep.

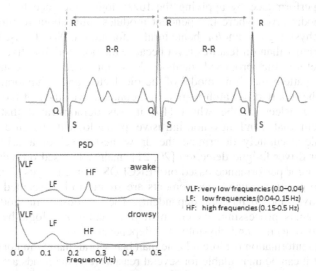

Fig. 1. PSD analysis of HRV is useful to understand when sleep is setting in

The HRV-based approach to the early detection of driver's fatigue and drowsiness is an effective alternative to the available implementations because it is fully noninvasive. The acquisition of the ECG signal from the hands works well both in collaborative and not collaborative automotive applications. The HRV signal is very rich of information concerning physiological, behavioral, and psychological state of the car driver, so it may not require additional information to infer about it. Anyway an extension of the HRV-based approach, holding its noninvasive peculiarity, is the integration of the breathing rate and of its dynamics with the purpose to strength the robustness of the inferring process. The breathing signal can be captured from seatbelt, featured, and fused with the HRV data. The breathing signal can be derived also from the ECG [10]. Felblinger [11] demonstrated that the envelope of the R peaks corresponds to the breathing signal, so our HRV-based approach in detecting drowsiness and fatigue is effective. Breathing and HRV are two independent features of the ECG because the first is its amplitude modulation of the ECG and the second is its frequency modulation, so they need to be fused at decision level.

Making inferences about physiological status from the HRV signal is very difficult, because of the high degree of variability and the presence of artifacts. Soft-computing methods and decision fusion can be very effective at making inferences in such a context [12]. There are several methods [13, 14, 15] for performing predictions with neural networks. Mager [15] uses Kohonen's self-organizing map (SOM) to provide a method of clustering subjects with similar features. This method, applied to detecting sleep onset early, allows artifacts to be filtered and the variability component of noise to be combined with the primary HRV signal ready for smoothing. An alternative approach uses fuzzy decision logic [16] to model sleep onset. Such an approach is effective, because it enables us to use membership functions to model data features.

Some research works [17, 18] demonstrate that fuzzy logic methods can be very effective if tailored on the specific nature of drowsiness detection and if ad hoc methods for fuzzification and setting the rules are implemented. The proposed method improves the performance by applying the fuzzy logic inference to fuse decisions from a multimodal layer where independent modules infer about features measured from sensed physiological and/or behavioral information. We focused mainly on decision level rather than on features level because the non-collaborative nature of the process. Therefore, the proposed method does not need of a complex signal processing and pattern matching model of the bioelectric and behavioral information as in most of the reported research works. In a comparative study of the methods for detecting falling asleep at the wheel [19] it was demonstrated that drowsiness detection system that combines non-intrusive physiological measures with other measures would accurately determine the drowsiness level of a driver. Methods investigated for driver fatigue detection [20, 21] mainly focused on measures of the driver's state and/or performance based on PERCLOS (Percent Eyes Closure), mouth shape and head position. Such measurements are simple to be executed but they are highly subjective and behavioral-dependent. The proposed method, based on bioelectrical signals processing, pattern matching, and fuzzy logic-based decision fusion, is mostly objective and physiological-dependent.

Physiological information is more reliable than behavioral, mainly in not collaborative application, but it can be unavailable for several reasons (e.g. the hands are not touching the steering wheel at the same time). Behavioral information is always available, but it depends on the car driver collaboration (e.g. avoiding using reflective sun glasses in slow eyes closure rate measurement). Calibration in visual methods is subject-dependent [22] and it needs to be executed prior to eye-state monitoring.

Relying on only one predictor of driver drowsiness makes the system susceptible of data unavailability due to failure of the single sensor or driver's individual differences [23]. Multi-measure approach enables data fusion methods to be applied so that more robust measures can be executed in field application. The cost of this approach is the increase of the complexity of the system and of the data fusion engine. To keep complexity low and boost the performance of the whole system, data fusion can be applied to decision level rather than feature level. The generalization capability of our system stays in its strategy based on fuzzy fusion at decision layer (upper) rather than at feature layer (lower). This enables the modules of the feature layer to work independently of each other on physiological and/or behavioral information and to be considered at decision layer only if they are meaningful.

2   System Framework

The system consists of two layers (Fig. 2). The lower layer has subsystems for capturing, detecting, and deciding locally, each operating on a single physiology or behavior signal. Each subsystem captures, conditions, preprocesses, extracts features, and makes a fuzzy logic-based decision. The upper layer fuses the fuzzy decisions delivered by each of the subsystems active at the lower layer.

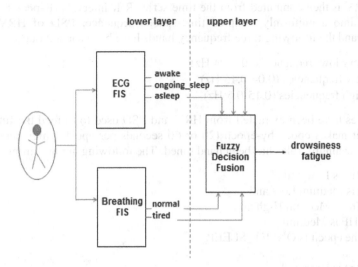

Fig. 2. The system consists of two layers: the lower layer has subsystems for capturing, detecting, and deciding locally; the upper layer fuses the fuzzy decisions delivered by each of the subsystems active at the lower layer

The fuzzy logic engine is a Mamdani fuzzy inference system using min and max for T-norm and T-conorm, and triangular, trapezoidal, and singletone membership functions. Five linguistics have been applied for inputs and three for outputs. Manual tuning of the rule set has been applied to solve the combinatorial explosion problem.

Natural extraction of membership functions strategy has been applied using an expert in the field of application to draw the membership curve according to the requirements. Using a graphical user interface (GUI) the expert chose among a restricted set of membership functions. The same strategy has been applied to compile the rules, using a predefined rule format.

A simulator has been used to hand tune the rules moving graphically the crisp inputs. All the rules have been evaluated and the fuzzy output inferred. The expert tunes the memberships and the rules looking to the input and output data.

2.1   ECG Subsystem

The ECG captured signal has been sampled and preprocessed to remove baseline fluctuations, muscle noise, and artifacts. Baseline oscillations are removed using a zero-phase, fourth-order, high-pass filter (1-Hz cutoff frequency).

To compute the R-R tacogram, the QRS complex of the ECG cycle a threshold has been applied. This was done by squaring the sample values and passing them through a moving average filter:

$$y(n) = \frac{1}{N} \sum_{i=0}^{N-1} x(n-i) \tag{1}$$

The HRV is then computed from the time series R-R intervals (R-peak to R-peak) converted into a uniformly sampled time-spaced sequence. PSD of HRV is then computed and the following three frequency bands have been carried out:

- very low frequencies (0-0.04 Hz)
- low frequencies (0.04-0.15 Hz)
- high frequencies (0.15-0.5 Hz)

The features have been extracted from HRV and PSD used to to feed the fuzzy logic engine that makes epoch-by-epoch (20 or 60 seconds per epoch) inferences [24]. A set of 23 meaningful rules has been hand tuned. The following are the strongest:

if HRV(n) is Low and
　LF(n) is Medium Low and
　　HF(n) is Medium High and
　　LF/HF is Medium
　then the epoch is ONSET_SLEEP
…
if HRV(n) is High and
　LF(n) is High and
　　HF(n) is Low and
　　LF/HF is High
　then the epoch is WAKE
…
if HRV(n) is Low and
　LF(n) is Low and
　　HF(n) is High and
　　LF/HF is Low
　then the epoch is SLEEP

2.2　Breathing Subsystem

The breathing subsystem measures the breathing signal captured using a MEMS (Micro Electro-Mechanical System) accelerometer in noninvasive contact with the driver's body (thorax). The breathing signal is low-pass filtered to remove high-frequency noise with the following algorithm:

$$
\begin{aligned}
&y(n) = ax(n) + by(n-1) \\
&a = 1-\exp(-2\pi fc/fs) \\
&b = 1-a \\
&fc: 1.0\ Hz \qquad // \text{ filter cutoff frequency} \\
&fs: 50\ Hz \qquad // \text{ sampling rate frequency}
\end{aligned}
\tag{2}
$$

Then, a high-pass filter is applied to remove very low frequencies and the baseline that conditions the captured breathing signal:

$$
\begin{aligned}
&y(n) = x(n) - R \\
&R = ax(n) + bR \\
&fc: 0.1\ Hz \qquad // \text{ filter cutoff frequency} \\
&fs: 50\ Hz \qquad // \text{ sampling rate frequency}
\end{aligned}
\tag{3}
$$

Breathing rate and amplitude are then measured and fuzzy-processed by a Mamdani-type fuzzy logic engine that, epoch by epoch, obtain inferences about the oncoming drowsiness or fatigue. A set of 19 rules has been hand tuned. The following are the strongest rules:

...

> if BreathingRate(n) is Medium and
>   BreathingAmplitude(n) is Medium
>   then the epoch is NORMAL
>
> ...   (4)
>
> if BreathingRate(n) is Low and
>   BreathingAmplitude(n) is High
>   then the epoch is TIRED
>
> ...

The center-of-gravity method (5) and the singleton membership function are then applied to defuzzify the decision locally:

$$crisp\_output = \frac{\sum (fuzzy\_output)\ (singleton\_position)}{\sum fuzzy\_output} \qquad (5)$$

2.3   Decision-Fusion Subsystem

Combining classifiers to make a decision demonstrates to be more efficient than compound classification [25]. Decision-fusion methods implement the decision level by integrating decisions from the ECG and the breathing subsystems. To improve identification of fatigue or drowsiness, a soft decision method was applied, because it proved superior to hard decision methods. The soft decision method consists of a set of 15 hand tuned fuzzy rules. The following are the strongest:

> ...
>
> if ECG(n) is Ongoing_Sleep and
>   Breathing(n) is Tired and
>   Hand_Movements(n) is Low
>   then the epoch is DROSINESS
>
> ...   (6)
>
> if ECG(n) is Ongoing_Sleep and
>   Breathing(n)_is Normal and
>   Hand_Movements(n) is Medium
>   then the epoch is FATIGUE

An additional input (Hand_Movements) has been considered at fusion level to demonstrate the generalization capability of the system. The additional input is the score from a lower module (not in the basic framework) that evaluates the movements of the hands by an accelerometer sensor. This module is considered as an example of add-on that can improve the whole system and/or replace a faulty or unavailable module.

The center-of-gravity method (5) and the singleton membership function are then applied to defuzzify the decision locally.

To overcome the limitation of the crisp output, a degree of sleepiness has been applied as consequent in the rules. This solution keeps low the complexity of the defuzzification stage and enables to decide about the degree of alarm to be activated.

3 Experimental Results

Three levels of tests were conducted, one on clinical captured signals (invasive) and two in a simulated driving environment. The first of the latter two tests was partially invasive and visible, because it used standard electrodes to capture the ECG and a laboratory biomedical instrument to collect and process data. The second of the latter two tests was fully noninvasive, because the entire system was integrated into an embedded device.

3.1 Embedded Prototype

The embedded prototype (Fig. 3) is based on a digital subsystem, a microcontroller unit (MCU), and an analog subsystem, an analog front-end (AFE). The MCU is the STM32 from STMicroelectronics, a 32 bit ARM M3-based computing architecture, integrated in the palm top computer Primer 2 from Raisonance that includes the analog and the digital interface, the accelerometer, the mass memory storage, and the human-machine interface (HMI) . The AFE is the AD8232 from Analog Devices, a single chip device that integrates all the analog resources requested to interface end-to-end the human body to the MCU for ECG signal acquisition. All the application has been ANSI-C encoded and flashed on the non-volatile memory of the MCU. A couple of stain steel electrodes has been directly connected the differential analog input of the AFE, considering the Right Leg (RL) electrode input common to one of the two analog differential inputs.

Fig. 3. Protosystem with a microcontroller from STMicroelectronics (STM32 ARM M3) and the analog front-end (AFE) from Analog Devices (AD8232)

3.2   Performance Evaluation

The results of both experiments confirmed our hypothesis that sleep onset can be predicted by using only features extracted from the HRV and that breathing information can improve the sleep-onset. Detection may be successfully based on ECG signal captured from the driver's hands and on breathing signal.

The results of the experiments (Table 1) mean that the decision fusion layer performs always better than the feature fusion layer. When the sensed signal is captured in an uncontrolled environment (embedded) and with not invasive methods, the decision fusion layer performs at least like the best subsystem of the feature fusion layer. Measurement accuracy is lower than the clinical context, above all for ECG module, but decision fusion layer confirms to be robust enough to perform not under 85% successful early detections.

Table 1. Matrix of confusion (a) and precision (b) for the three levels of tests conducted, one on clinical captured signals (invasive) and two (noninvasive) in a simulated driving environment

(a)

	Decision fusion		ECG features fusion		Breathing features fusion	
	PREDICTED					
	Drosiness	Normal	Drosiness	Normal	Drosiness	Normal
ACTUAL						
Clinical						
Drosiness	19	1	19	1	18	2
Normal	1	19	3	17	2	18
In the field (nonembedded)						
Drosiness	18	2	18	2	17	3
Normal	2	18	4	16	1	19
In the field (embedded)						
Drosiness	17	3	17	3	16	4
Normal	3	17	5	15	2	18

(b)

Test type	Decision fusion	ECG features fusion	Breathing features fusion
Clinical	95%	90%	90%
In the field (nonembedded)	90%	85%	90%
In the field (embedded)	85%	80%	85%

The performance of each system has been validated by a human (expert). For clinical tests, a physician reads the bioelectric data and evaluates, for each module, if the identified transition from awake to asleep is acceptable or not. For in the fied tests this was done by visual observation of the awake/asleep status of the subject and the consequent qualification of the detection done by each system of the transition awake/asleep. Timing of the early detection capability of each system was also evaluated during the tests.

4   Conclusions and Future Work

The proposed system for early detection of driver fatigue and drowsiness is not directly comparable to the systems currently implemented or under investigation, due to its different system architecture and methodology. Most of the systems are based on a single methodology and are mainly behavioral [19].

Tests of behavioral-based systems (mainly visual) performed from 85% to 100% successful detection [19]. Most of these are experiments conducted in simulated environment. These success rates decreased significantly when the tests were carried out in the field. It is also important to consider that visual-based systems need of the driver's collaboration (e.g. is requested that the subjects don't wear glasses). An important consideration is also that successful detections in behavioral-based systems can be effective only after the driver starts to sleep, too late to prevent the crash.

Systems based on the measurement of the physiologic features is performing well like the behavioral in classification accuracy (typically 90%) and do not require the subject collaboration, but it can be very sensitive to artifacts [27].

Improvements can be gained with more effective methods for feature extraction. The cardiac vagal index (CVI) and the cardiac sympathetic index (CSI) have been found to be more reliable than those obtained by the other methods [28].

4.1   Adaptive Fuzzy Featuring

The HRV and the breathing fuzzy featuring layers are both modeled using the crisp data at design time. This is useful to setup the fuzzy inferential engine to fulfill the inference target, but it is also useful to build up an incremental learning capability that enables an embedded evolving ability for the whole system.

Adaptation is a strong requirement because the application is unattended and the in the field operability of the crisp features modeled at design time can change, so the membership functions and the rules need to be tuned. If some crisp measurements vary significantly, then a new tuning action needs to starts on the membership functions and the rule set.

To implement the adaptation capability, manually tuning of the fuzzy engine is not a practical solution. Next step in the development of the fuzzy engine of this system will apply the EFuNNs (Evolving Fuzzy Neural Networks), according to the ECOS framework (Evolving COnnectionist Systems) [26]. This algorithm treats each evolving layer neuron as a fuzzy rule, and finds the connections with the largest weights. The connection weights in EFuNN represent also the fuzzified input and output vectors, so selecting the winning weights, the algorithm is finding the MFs that best fit.

References

1. NHTSA: Drowsy driving. Published by NHTSA's national center for statistics and analysis 1200 New Jersey Avenue SE., Washington, DC 20590 (2011)
2. Eriksson, M., Papanikolopoulos, N.P.: Eye-tracking for Detection of Driver Fatigue. In: IEEE Proceendings of Intelligent Transport System, Boston, MA, pp. 314–319 (1997)
3. Malcangi, M., Smirne, S.: Fuzzy-logic inference for early detection of sleep onset in car driver. In: Jayne, C., Yue, S., Iliadis, L. (eds.) EANN 2012. CCIS, vol. 311, pp. 41–50. Springer, Heidelberg (2012)
4. Dorfman, G.F., Baharav, A., Cahan, C., Akselrod, S.: Early Detection of Falling Asleep at the Wheel: a Heart Rate Variability Approach. Computers in Cardiology 35, 1109–1112 (2008)
5. Zocchi, C., Giusti, A., Adami, A., Scaramellini, F., Rovetta, A.: Biorobotic system for increasing automotive safety. In: 12th IFToMM World Congress, Besançon, France (2007)
6. Estrada, E., Nazeran, H.: EEG and HRV Signal Features for Automatic Sleep Staging and Apnea Detection. In: 20th International Conference on Electronics, Communications and Computer (CONIELECOMP), February 22-24, pp. 142–147 (2010)
7. Manis, G., Nikolopoulos, S., Alexandridi, A.: Prediction techniques and HRV analysis. In: MEDICON 2004, Naples, Italy, July 31-August 5 (2004)
8. Rajendra, A.U., Paul, J.K., Kannathal, N., Lim, C.M., Suri, J.S.: Heart rate variability: a review. Med. Bio. Eng. Comput. 44, 1031–1051 (2006)
9. Tohara, T., Katayama, M., Takajyo, A., Inoue, K., Shirakawa, S., Kitado, M., Takahashi, T., Nishimur, Y.: Time frequency analysis of biological signal during sleep. In: SICE Annual Conference, September 17-20, pp. 1925–1929. Kagawa University, Japan (2007)
10. Travaglini, A., Lamberti, C., DeBie, J., Ferri, M.: Respiratory signal derived from eight-lead ECG. Computer in Cardiology 25, 65–68 (1998)
11. Felblinger, J., Boesch, C.: Amplitude demodulation of the electrocardiogram signal (ECG) for respiration monitoring and compensation during MR examinations. Magn-Reson-Med. 38(1), 129–136 (1997)
12. Patel, M., Lal, S.K.L., Kavanagh, D., Rossiter, P.: Applying neural networks analysis on heart rate variability data to asses driver fatigue. Expert systems with Applications (2011)
13. Ranganathan, G., Rangarajan, R., Bindhu, V.: Signal processing of heart rate variability using wavelet transform for mental stress measurement. Journal of Theoretical and Applied Information Technology 11(2), 124–129 (2010)
14. Ranganathan, G., Rangarajan, R., Bindhu, V.: Evaluation of ECG signal for mental stress assessment using fuzzy technique. International Journal of Soft Computing and Engineering (IJSCE) 1(4), 195–201 (2011)
15. Mager, D.E., Merritt, M.M., Kasturi, J., Witkin, L.R., Urdiqui-Macdonald, M., Sollers, J.I., Evans, M.K., Zonderman, A.B., Abernethy, D.R., Thayer, J.F.: Kullback–Leibler Clustering of Continuous Wavelet Transform Measures of Heart Rate Variability. Biomed. Sci. Instrum. 40, 337–342 (2004)
16. Dzitac, S., Popper, L., Secui, C.D., Vesselenyi, T., Moga, I.: Fuzzy Algorithm for Human Drowsiness Detection Devices. SIC 19(4), 419–426 (2010)
17. Sharma, N., Banga, V.K.: Development of a drowsiness warning system based on the fuzzy logic. International Journal of Computer Applications (0975-8887) 8(9) (2010)
18. Picot, A., Charboinner, S., Caplier, A.: Drowsiness detection based on visual signs: blinking analysis based on high frame rate video. In: 2010 IEEE International Instrumentation and Measurement Technology Conference, 2MTC 2010 (2010)

19. Sahayadhas, A., Sundaraj, K., Murugappan, M.: Detecting driver drowsiness based on sensors: A review. Sensors 2012 12, 16937–16953 (2012)
20. Wang, Q., Yang, J., Ren, M., Zheng, Y.: Driver fatigue detection: a survey. In: Proceedings of the 6th World Congress of Intelligent Control and Automation, pp. 8587–8591. IEEE (2006)
21. Bajaj, P., Narole, N., Devi, M.S.: Research on Driver's Fatigue Detection. eNewsletter System, Man and Cybernetics Society (31) (June 2010)
22. Albu, A.B., Widsten, B., Wang, T., Lan, J., Mah, J.: A Computer Vision-based System for Real-time Detection of Sleep Onset in Fatigued Drivers. In: Proceedings of 2008 IEEE Intelligent Vehicles Symposium, Eindhoven University of Technology Eindhoven, The Netherlands, June 4-6, pp. 25–30 (2008)
23. Bowman, D.S., Schaudt, W.A., Hanowski, R.J.: Advances in Drowsy Driver Assistance Systems through Data Fusion. In: Handbook of Intelligent Vehicles, pp. 895–912. Springer (2012)
24. Malcangi, M., Smirne, S.: Heart Rate Variability Analysis for Prediction of Sleep Onset in Car Drivers. Journal of Sleep Research 21(Suppl. 1), 307–308 (2012)
25. Kittler, J., Hatef, M., Duin, R.P.W., Matas, J.: On cobining classifier. IEEE Transactions on Pattern Analysis and Mahine Intelligence 20(3), 226–239 (1998)
26. Kasabov, N.: Evolving fuzzy neural networks – algorithms, applications and biological motivation. In: Yamakawa, Matsumoto (eds.) Methodologies for the conception, design and application of the soft computing, World Computing, pp. 271–274 (1998)
27. Sandberg, D., Anund, A., Fors, C., Kecklund, G., Karlsson, J.G., Wahde, M., Åkerstedt, T.: The characteristics of sleepiness during real driving at night—A study of driving performance, physiology and subjective experience. Sleep 34(10), 1317–1325 (2011)
28. Lin, C.W., Wang, J.S., Chung, P.C.: Mining Physiological Conditions from Heart Rate Variability Analysis. IEEE Computational Intelligence Magazine, 50–58 (2010)

Neural Trade-Offs among Specialist and Generalist Neurons in Pattern Recognition

Aarón Montero[1], Ramón Huerta[1,2], and Francisco B. Rodríguez[1]

[1] Grupo de Neurocomputación Biológica, Dpto. de Ingeniería Informática. Escuela Politécnica Superior. Universidad Autónoma de Madrid, 28049 Madrid, Spain
[2] BioCircuits Institute, University of California, San Diego, La Jolla, CA 92093-0402, USA
aaron.montero@uam.es

Abstract. The olfactory system of insects has two types of neurons based on the conditional response to odorants. Neurons that respond to a few odor classes are called specialists, while generalist neurons code for a wide range of input classes. The function of these neurons is intriguing. Specialist neurons are perhaps essential for odor discrimination, while generalist neurons may extract general properties of the odor space to be able to generalize to new odor spaces. Our goal is to shed light on this issue by analyzing the relevance of these neurons for pattern recognition purposes. The computational model is based on the olfactory system of insects. The model contains an approximation to the antennal lobe (AL) and mushroom body (MB) using a single-hidden-layer neural network. To determine the optimal balance between specialists and generalists we measure the classification error of the pattern recognition task. The mechanism to achieve the optimal balance is synaptic pruning to select the optimal synaptic configuration. The results show that specialists play an important role in odor classification, which is not observed for generalists. Furthermore, proper classification requires low neural activity in Kenyon cells, KC, which is consistent with the sparseness condition observed in MB neurons. Moreover, we also observe that the model is robust against noise to input patterns showing better resilience for low connection probabilities between AL and MB.

Keywords: Pattern recognition, generalist neuron, specialist neuron, olfactory system, neural variability, synaptic pruning, supervised learning, heterogeneous threshold.

1 Introduction

Specialist neurons are selective responding to stimuli, while generalists code for multiple stimuli. The role of both classes of neurons in the olfactory system is still under debate [14,4]. However, it is suggested that specialist neurons are crucial for discrimination, while generalist neurons play a key role in extracting and discovering common features [27]. In order to assess the role of specialist and generalist neurons in pattern recognition performance, we use neural sensitivity.

V. Mladenov et al. (Eds.): EANN 2014, CCIS 459, pp. 71–80, 2014.

It can be estimated from the distribution of neurons that respond to n out of N stimuli [20,21]. This allows to define neurons as specialists or generalists. However, because the boundary between specialists is arbitrary in a continuous distribution of sensitivity a systematic analysis is required on this distribution. The question we want to answer is the ratio of neural types that lead to best pattern recognition in test sets.

To answer this question, we chose the olfactory system of insects because of the presence of specialist and generalist neurons [18,4,20,28,21]. The insect model also has a well-defined structural organization [7,13] that allows quick and stable odorant discrimination [10] and there is extensive literature on the dynamics of learning during discrimination tasks [3,24].

In insects, an odor is intercepted by the antenna, where a massive number of receptors encode its stimulus in a high-dimensional space. This information is received by the antennal lobe, AL. The AL output is received by a wide number of Kennyon cells, KCs, of the mushroom body using a fan-out connectivity that increases the separability between different odor encodings. This fan-out phase combined with the sparse firing for these KCs [18,8] facilitates the odorant discrimination process realized in following fan-in phase by output neurons, which are involved in memory formation and storage.

We used a computational model focused on the AL and MB, using a single-hidden-layer neural network. The input of these neural network is the AL activity, that is connected to MB through a non-specific connectivity matrix [8,15]. The other layers, hidden and output, are made of KCs and output neurons respectively. These neurons are connected by a connectivity matrix subjected to learning that is modulated by Hebbian learning [2]. Moreover, neuronal thresholds used are heterogeneous, since it was observed that there is threshold variability in neuronal populations of the olfactory system [18,11] and these achieve a classification enhancement in the artificial noses [5] and computer models [16]. These thresholds also allow a greater variability in neuronal sensitivity and therefore a better differentiation between responses of neurons (specialists and generalists).

Our focus is on studying the importance of specialist and generalist neurons for pattern recognition performance. For this purpose, we use three different strategies of synaptic pruning based on sensitivity properties of neurons. The synaptic pruning is a neurological regulatory process, which allows to selectively remove exuberant neuronal branches and connections in the immature nervous system to ensure the proper formation of functional circuitry [6,19]. Furthermore, we use different degrees of overlapping patterns to study their effects over odor classification and neural balance.

We will show in results that when only the most specialist KCs are able to fire and transmit their information, the classification error improves. This implies that the generalist neurons do not influence over odor classification process and only the most specialists are really relevant for this purpose. Furthermore, classification errors are lower when the connection probability between AL and MB is low as well.

2 Methods

In this section, we provide the structure of the computational model of the olfactory system as well as the pattern learning associated with it. We also describe the patterns introduced in the model and methods for investigating the relevance of specialist and generalist neurons. These methods comprise threshold selection and synaptic pruning. The first method employs heterogeneous thresholds to facilitate differentiation of neurons by neural sensitivity. The second one selects neurons with different sensitivity degrees (specialists and generalists) and analyze its performance for the pattern recognition.

2.1 Model

The model focuses on the AL and MB, dividing the MB into KCs and output neurons. Therefore, the network model is a single-hidden-layer neural network with an input layer of 50 neurons, a hidden layer with 2,500 neurons (locust has a ratio of 1:50 between neurons of the AL, input layer, and KCs, hidden layer) and an output layer with 5 neurons [12]. These dimensions were chosen because they ensure a high probability of classification for the input used [11] for a relatively low computational cost.

The KC neurons of the MB display very low activity [18]. These neurons are inactive most of the time, with a mean firing frequency lower than 1 Hz. But when they are activated, their neuronal response is produced by the coincidence of concurrent spikes followed by a reset. Bearing in mind this behavior, we chose the McCulloch-Pitts model in all neurons of the hidden and output layers. Therefore, we have the following:

$$y_j = \varphi(\sum_{i=1}^{N_{AL}} c_{ji} x_i - \theta_j), \quad j = 1, \ldots, N_{KC},$$

$$z_l = \varphi(\sum_{j=1}^{N_{KC}} w_{lj} y_j - \varepsilon_l), \quad l = 1, \ldots, N_{OutN},$$

where x_i, y_j and z_l are activation states for a input, hidden and output neuron respectively. The weights c_{ji} and w_{lj} link two neurons, for input and hidden layer in case of matrix C, for hidden and output layer in case of matrix W. The thresholds for the hidden and output layer are θ_j and ε_l respectively. The Heaviside activation function φ is 0 when its argument is negative or 0 and 1 otherwise.

The connectivity matrices, C and W, are initialized at the beginning of each learning process. These connectivities are created by using the connection probabilities, p_c and p_w, as a threshold on matrices with random values uniformly distributed. The connectivity matrix W is updated using Hebbian learning, while matrix C remains fixed.

The synaptic model is binary except for the input neurons, that are real numbers (see subsection: Patterns). Therefore, activation states for a MB neuron and weights can only take values 0 or 1.

2.2 Hebbian Learning

The connectivity matrix W connects the KCs and output neurons. Their connections have an associative learning, that can be simulated by using Hebbian learning [2]. This learning is subjected to a target t of the output layer (supervised learning) and certain thresholds, whose selection will be detailed later (see subsection: Threshold selection). Hebbian learning allows the strengthening or weakening the connections of a connectivity matrix, as follows [11,12]:

$$w_{lj}(n+1) = H(t_l, y_j, w_{lj}(n)),$$

$$H(1,1,w_{lj}(n)) = \left\{ \begin{array}{ll} 1 & \textit{with probability } p_+, \\ w_{lj}(n) & \textit{with probability } 1 - p_+, \end{array} \right.$$

$$H(1,0,w_{lj}(n)) = \left\{ \begin{array}{ll} 0 & \textit{with probability } p_-, \\ w_{lj}(n) & \textit{with probability } 1 - p_-, \end{array} \right.$$

$$H(0,1,w_{lj}(n)) = w_{lj}(n), \quad H(0,0,w_{lj}(n)) = w_{lj}(n),$$

where the future connection state $w_{lj}(n+1)$ is determined by a function H. This function $H(t_l, y_j, w_{lj}(n))$ depends on the target for the output layer neuron t_l, the hidden layer neuron y_j and the current connection state $w_{lj}(n)$. If the target for the output layer neuron does not fire, the connection state is not changed. However, if the target neuron fires, the connection state depends on the hidden layer in the following ways: i.If the hidden layer neuron has fired, then the connection between these neurons is created with a probability p_+. ii.If the hidden layer neuron has not fired, then the connection between these neurons is destroyed with a probability p_-.

2.3 Patterns

The AL forms a code of an odor space [22]. Different levels of odor concentration can expand the number of neurons that are activated for a certain odorant. This can increase the overlapping with the regions used by other odorants [22]. Also, recordings from the AL in the locust indicate that its activity remains nearly constant despite large variations of the odor concentration [25]. Therefore, a gain control mechanism [23] controlling neuronal activity in the AL is likely to exist.

Considering this, we used Gaussian patterns centered at different input neurons depending on the class to which they belong (Fig. 1, where we can see a pattern example for each of their classes and configurations) . These patterns represent the spikes probabilities in AL neurons, that can involve a variable number of neurons. Number of neurons that we controlled by standard deviation and implies different overlapping degrees. The standard deviations, overlapping degrees, used in our experiments are: 1, 2.5, 5, 10, 25. Furthermore, we also added noise to the patterns, so that we can observe the robustness of the model and

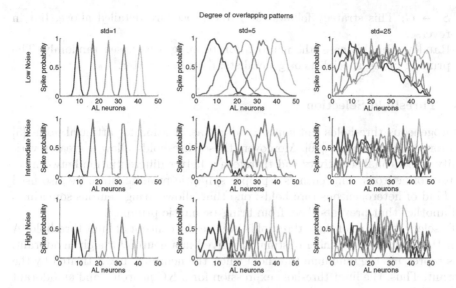

Fig. 1. Examples of patterns to classify. Each panel shows an input example for each of the 5 classes. The patterns vary depending on noise (rows) and overlapping degrees (columns). Although for the study we used five standard deviations for the overlapping degrees, in this figure we only show three of them: 1, 5, 25. The X axis represents the 50 neurons of the AL, while the Y axis represents their spike probability.

its impact on odor classification. We have three different noise degrees: low, intermediate and high. Therefore, we have 15 pattern configurations (overlapping and noise degrees) and 5 classes of them. We used 100 patterns for each pattern configuration, 20 patterns for each class. In the learning process these patterns are divided into 5 parts, taking one as test set an the other four as training set. This process is repeated 5 times, in order to each part can be used as test set. Thus, the training data set has 80 patterns and test set has 20 patterns.

2.4   Synaptic Pruning Strategies

To analyze the relevance of specialist and generalist neurons for odor classification, we have designed different synaptic pruning strategies. Synaptic pruning refers to the process that eliminates excessive or inappropriate synapses to form proper synaptic connections during development of neurons [6,19]. We use different values of neural sensitivity to determine the relevance of hidden neurons [1,9] and therefore the neurons that have to be pruned. These pruning allow us to observe that odor classification is obtained when only neurons with a certain odor sensitivity are able to transmit information to the output layer.

The strategies that we have designed for this study are:

- $G \rightarrow S$: In this modality, we perform pruning by neural sensitivity, starting with the most generalist neurons and ending with the most specialist ones.

- $S \rightarrow G$: This strategy follows the same procedure detailed above, but in reverse.
- Random: In this case, the neurons for pruning are chosen randomly. The pruning order, specialists or generalists, does not matter.

2.5 Threshold Selection

Heterogeneous thresholds can improve odor classification for artificial noses [5] olfactory system models [16]. Moreover, these thresholds allow a greater variability in neuronal sensitivity and therefore a better differentiation between responses of neurons. This ensures a proper synaptic pruning, since we also used two kind of heterogeneous thresholds: one that allows firing neurons sometimes and another that prevents them from firing (synaptic pruning).

To select the value of these thresholds, we use the concept of limit threshold. A limit threshold is the number of stimuli received in a neuron for a given odorant. This represents the minimum threshold when the neuron is not activated by the odorant. Thus, the limit threshold expression for a KC neuron j and an odorant O is as follows:

$$\theta_j^O = \sum_{i=1}^{N_{AL}} c_{ji} x_i^O,$$

where neuron j spikes $\forall \theta_j, 0 \leq \theta_j < \theta_j^O$.

For synaptic pruning, we take the maximum limit threshold for each neuron. These thresholds prevent neural spikes, at least for the training patterns. For the remaining thresholds, we choose the limit threshold, for each neuron, in order to make the neuron as specialist as possible. This is because many neurons exhibit a bimodal distribution and therefore a predilection for certain odorants.

We calculate the limit threshold matrix for the hidden layer and the neural sensitivity for their selected thresholds before Hebbian learning is applied.

3   Results

The following averaged results for the test set were obtained by supervised learning. This process begins with the division of the patterns as explained in Patterns subsection. Then, we selected the thresholds that make specialist neurons for the training set. The sensitivity of neurons is obtained for these thresholds and is used to sort the neurons. To determine the relevance of specialists and generalists neurons in the framework of odor classification, we used three synaptic pruning strategies by using this sensitivity order (see subsection: Synaptic pruning strategies).

We used 45 system configurations, one for each combination of the parameters of the study: connection probability (3), overlapping (5) and noise (3); and we ran 10 simulations for each of them. We have used p_c connection probabilities with values from 0.1 to 0.5 based on previous studies [8]. The p_w value is 0.5

because of its matrix is subjected to Hebbian learning [12]. The combination of values for Hebbian probabilities that optimize the result are $p_+ = 0.2$ and $p_- = 0.1$, from a wide range of analyzed combination, $[0, 1] \in \mathbb{N}$ [16,17].

3.1   Relevant Role of Specialists Neurons for Odor Classification

These results (Fig. 2) show that the best synaptic pruning starts with the most generalist neurons to end with the most specialist ones, $G \to S$. This is the unique pruning strategy that minimizes the classification error. This implies that to achieve minimum classification error the specialists are only required.

The firing rates (%) for KC neurons, observed for those points where the error is minimized, indicate that minimum error is found for a sparse activity. These results are consistent with the sparseness condition observed in these MB neurons. Also, we see that the model is more robust to noise for low connectivity probabilities, $p_c = 0.1$, consistent with other research [8,16,17].

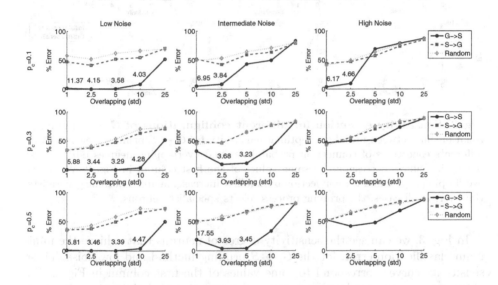

Fig. 2. Classification error for different synaptic pruning and connection probabilities. Comparison between different synaptic pruning strategies for test sets of odorants (5 classes), for different degrees of overlapping patterns and connection probabilities. The values shown in some points of the error values, represent the firing rates (%) for KC neurons where the error is minimized. For example, for low noise, connection probability $p_c = 0.5$ and overlapping degree $std = 5$, the firing rate is 3.39%. The $G \to S$ pruning achieve the best classification error in all cases. These results show that the model is more robust to noise for low connectivity probabilities.

3.2 From an Initial Generalist Sensitivity to a Specialist Sensitivity

Once known odor classification for different synaptic pruning strategies, we wonder whether neural sensitivity and percentage of active neurons are related to minimum classification error.

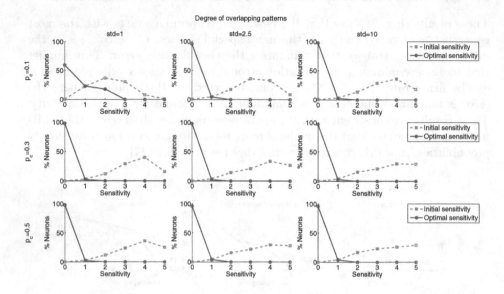

Fig. 3. Sensitivity evolution for different configurations of $G \to S$ pruning. Relationship between initial and optimal sensitivity for $G \to S$ pruning, low noise and differents conditions of connection probabilities and overlapping degrees (std). These sensitivity curves correspond to some values of the first column in Fig. 2. When the overlapping degree and the connection probability increase, neurons become generalists. However, in all cases, the pruning process selects specialist neurons.

In Fig. 3, we can see the sensitivity degree of neurons that achieve the minimum classification error, for the $G \to S$ pruning method and low noise. These sensitivity curves correspond to some values of the first column in Fig. 2. We observe that when overlapping degree and connection probability increase, neurons become generalists. However, in all cases, the pruning process leaves just the specialist neurons and greatly reduces the percentage of active neurons (sensitivity 0), which minimize the classification error.

4 Conclusions

The objective of this work is to investigate what is the role of specialist and generalist neurons in the framework of odorant classification. We conclude that the specialists neurons are critical for classification performance and that generalist neurons need to be pruned or, at least, controlled. To investigate the classification performance we used a simple model that retains the most relevant structural properties of the olfactory system. This model focuses on the AL and MB, where the input to single-hidden-layer neural network is the AL activity. The other layers, hidden and output that represent the MB, are composed by KCs and output neurons respectively. These latter layers are connected by a connectivity matrix that implements a supervised Hebbian learning. Using patterns with different overlapping degrees, we compared neuron populations with different neural sensitivity through different synaptic pruning strategies. This process shows how important the specialists neurons are.

We show that to achieve minimum error classification only the specialist neurons are required. In this network configuration, the percentages of active neurons are remarkably low which is consistent with the sparseness condition in KCs [18,26]. Moreover, high noise conditions and connection probabilities, leads to a lack of appropriate neurons that minimize classification error. Furthermore, classification errors are lower when the connection probability between AL and MB, p_c is low as well, which is consistent with information maximization criteria provided in [8].

Acknowledgments. This work was supported by the Spanish Government project TIN2010-19607 and predoctoral research grant BES-2011-049274. R.H. acknowledges partial support by NIDCD-R01DC011422-01.

References

1. Augasta, M.G., Kathirvalavakumar, T.: A novel pruning algorithm for optimizing feedforward neural network of classification problems. Neural Process Lett. 34, 241–258 (2011)
2. Bazhenov, M., Huerta, R., Smith, B.H.: A computational framework for understanding decision making through integration of basic learning rules. The Journal of Neuroscience 33(13), 5686–5697 (2013)
3. Bitterman, M.E., Menzel, R., Fietz, A., Schäfer, S.: Classical conditioning of proboscis extension in honeybees (apis mellifera). J. Comp. Psychol. 97(2), 107–119 (1983)
4. Christensen, T.A.: Making scents out of spatial and temporal codes in specialist and generalist olfactory networks. Chem. Senses 30, 283–284 (2005)
5. Doleman, B.J., Lewis, N.S.: Comparison of odour detection thresholds and odour discriminablities of a conducting polymer composite electronic nose versus mammalian olfaction. Sensors and Actuators B 72, 41–50 (2001)
6. Meilijson, I., Chechick, G., Ruppin, E.: Neuronal regulation: A mechanism for synaptic pruning during brain maturation. Neural Comput. 11(8), 2061–2080 (1999)
7. Galizia, C.G., McIlwrath, S.L., Menzel, R.: A digital 3D atlas of the honeybee antennal lobe based on optical sections acquired using confocal micoscropy. Cell Tissue Res. 295, 383–394 (1999)
8. Garcia-Sanchez, M., Huerta, R.: Design parameters of the fan-out phase of sensory systems. J. Comput. Neurosci. 15, 5–17 (2003)

9. Tan, A., Zhu, Z., Rong, H., Ong, Y.: A fast pruned-extreme learning machine for classification problem. Neurocomputing 72, 359–366 (2008)
10. Huerta, R.: Learning pattern recognition and decision making in the insect brain. AIP Conference Proceedings 1510, 101 (2013)
11. Huerta, R., Nowotny, T., Garcia-Sanchez, M., Abarbanel, H.D.I., Rabinovich, M.I.: Learning classification in the olfactory system of insects. Neural Comput. 16, 1601–1640 (2004)
12. Huerta, R., Nowotny, T.: Fast and robust learning by reinforcement signals: Explorations in the insect brain. Neural Comput. 21, 2123–2151 (2009)
13. Ito, K., Suzuki, K., Estes, P., Ramaswami, M., Yamamoto, D., Strausfeld, N.J.: The organization of extrinsic neurons and their implications in the functional roles of the mushroom bodies in *Drosophila melanogaster Meigen*. Drosophila Melanogaster Meigen 5, 52–77 (1998)
14. Kaupp, U.B.: Olfactory signalling in vertebrates and insects: differences and commonalities. Nature Reviews Neuroscience 11, 188–200 (2010)
15. Marin, E.C., Jefferis, G.S., Komiyama, T., Zhu, H., Luo, L.: Representation of the glomerular olfactory map in the *Drosophila* brain. Cell 109, 243–255 (2002)
16. Montero, A., Huerta, R., Rodríguez, F.B.: Neuron threshold variability in an olfactory model improves odorant discrimination. In: Ferrández Vicente, J.M., Álvarez Sánchez, J.R., de la Paz López, F., Toledo Moreo, F. J. (eds.) IWINAC 2013, Part I. LNCS, vol. 7930, pp. 16–25. Springer, Heidelberg (2013)
17. Montero, A., Huerta, R., Rodriguez, F.B.: Regulation of specialists and generalists by neural variability improves pattern recognition performance. In: Neurocomputing (submitted 2014)
18. Perez-Orive, J., Mazor, O., Turner, G.C., Cassenaer, S., Wilson, R.I., Laurent, G.: Oscillations and sparsening of odor representations in the mushroom body. Science 297(5580), 359–365 (2002)
19. Reed, R.: Pruning algorithms - a survey. IEEE Transactions on Neural Networks 4(5), 740–747 (1993)
20. Rodríguez, F.B., Huerta, R.: Techniques for temporal detection of neural sensitivity to external stimulation. Biol. Cybern. 100(4), 289–297 (2009)
21. Rodríguez, F.B., Huerta, R., Aylwin, M.: Neural sensitivity to odorants in deprived and normal olfactory bulbs. PLoS ONE 8(4) (2013)
22. Rubin, J.E., Katz, L.C.: Optical imaging of odorant representations in the mammalian olfactory bulb. J. Neurophysiol. 23, 449–511 (1999)
23. Serrano, E., Nowotny, T., Levi, R., Smith, B.H., Huerta, R.: Gain control network conditions in early sensory coding. PLoS Computational Biology 9(7) (2013)
24. Smith, B.H., Wright, G.A., Daly, K.C.: Learning-based recognition and discrimination of floral odors. In: Dudareva, N., Pichersky, E. (eds.) Biology of Floral Scent, ch. 12, pp. 263–295. CRC Press (2005)
25. Stopfer, M., Jayaraman, V., Laurent, G.: Intensity versus identity coding in an olfactory system. Neuron 39, 991–1004 (2003)
26. Strube-Bloss, M.F., Nawrot, M.P., Menzel, R.: Mushroom body output neurons encode odor-reward associations. J. Neurosci. 31(8), 3129–3140 (2011)
27. Wilson, R.I., Turner, G.C., Laurent, G.: Transformation of olfactory representations in the drosophila antennal lobe. Science 303(5656), 366–370 (2004)
28. Zavada, A., Buckley, C.L., Martinez, D., Rospars, J.-P., Nowotny, T.: Competition-based model of pheromone component ratio detection in the moth. PLoS One 6(2), e16308 (2011)

Classification of Events in Switch Machines Using Bayes, Fuzzy Logic System and Neural Network

Eduardo Aguiar[1], Fernando Nogueira[1], Renan Amaral[1], Diego Fabri[2],
Sérgio Rossignoli[2], José Geraldo Ferreira[2], Marley Vellasco[3],
Ricardo Tanscheit[3], Moisés Ribeiro[1], and Pedro Vellasco[4]

[1] Federal University of Juiz de Fora
Industrial and Mechanical Engineering Department and Electrical Engineering
Post-Graduation Program, Juiz de Fora/MG, Brazil
[2] MRS Logística S.A.
Juiz de Fora/MG, Brazil
[3] Pontifical Catholic University of Rio de Janeiro
Electrical Engineering Department, Rio de Janeiro/RJ, Brazil
[4] State University of Rio de Janeiro
Civil Engineering Department, Rio de Janeiro/RJ, Brazil

Abstract. The Railroad Switch denotes a set of parts in concordance
with two lines in order to allow the passage of railway vehicles from
one line to another. The Switch Machines are equipments used for handling Railroad Switches. Among all possible defects that can occur in a
electromechanical Switch Machine, this work emphasizes the three main
ones: the defect related to lack of lubrication, the defect related to lack of
adjustment and the defect related to some component of Switch Machine.
In addition, this work includes the normal operation of these equipments.
The proposal in question makes use of real data provided by a company
of the railway sector. Observing these four events, it is proposed the use
of Signal Processing and Computational Intelligence techniques to classify the mentioned events, generating benefits that will be discussed and
thus providing solutions for the company to reach the top of operational
excellence.

Keywords: Classification, Switch Machine, Bayes, Fuzzy Logic System,
Neural Networks.

1 Introduction

This work proposes the development of a methodology based on Signal Processing and Computational Intelligence (CI) techniques for monitoring Switch Machine (SM), where the database is real and provided by the MRS Logística S.A.
(https://www.mrs.com.br/). Currently there is a multi-year project that aims
to increase the number of sensorized SMs in the field and attempts to cover 100
% of SMs from the company (currently there are 624 units). With the increase

V. Mladenov et al. (Eds.): EANN 2014, CCIS 459, pp. 81–91, 2014.

in the number of sensorized units, it becomes vitally important the improvement of fault detection systems, mostly with the focus on predictive maintenance.

Thus, this paper aims to compare the performance of three classifiers based on Bayes, Fuzzy Logic System and Neural Networks which will be the basis of a monitoring system based on CI, which, through the existing knowledge base in MRS Logística SA, will be able to classify failures through remote monitoring of the current from the motor of the SM present in the field. Therefore, this would make it possible to reduce the impact on the operation of trains and in the number of cyclical preventive maintenance, give the fact that interventions shall be carried out only when deviations are observed in the equipment, identifying potential points of failure and also indicate what type of failure occurred in the asset. The problem of classification be widely discussed in the literature, such as [1,2]. Furthermore, classification techniques has been applied on railway area in works like [3,4]. It is important to emphasize the absence of applications that aims to classify events in Switch Machines.

In this work it will be assessed the performance of three classifiers: Bayes Classifier [2], based on Maximum Likelihood (ML) criterion, other based on type-1 and singleton Fuzzy Logic System (FLS) [5] and finally one based on a Multilayer Perceptron (MLP) Neural Netork [6].

2   Problem Formulation

Let $\mathbf{x}$ a vector of sample signal with N elements. Figure 1 shows the paradigm used for the classification of events. The block "Feature Extraction" is responsible for extracting features (parameters), so that later there may be a selection of features. Note that $\mathbf{p}_l$, $\mathbf{p}_a$, $\mathbf{p}_c$ and $\mathbf{p}_n$ refer respectively to the vector of features extracted events called lubrication, adjustment, component and normal operation. Finally, after obtaining the feature vector, the block "Classification" applies one of the classification techniques to be presented to obtain the output vector $\mathbf{s}$, thereby deciding the type of present event in the input vector $\mathbf{x}$.

Each block and their characteristics will be presented in following sections. As a result, the classification of events in the component $\mathbf{x}$ can be formulated as a simple decision between hypotheses related to the occurrence of the events covered in this work, as shown below:

$$
\begin{aligned}
\mathcal{H}_{x,0} &: \mathbf{x} = \mathbf{x}_{lub}, \\
\mathcal{H}_{x,1} &: \mathbf{x} = \mathbf{x}_{adj}, \\
\mathcal{H}_{x,2} &: \mathbf{x} = \mathbf{x}_{comp}, \\
\mathcal{H}_{x,3} &: \mathbf{x} = \mathbf{x}_{norm}.
\end{aligned}
\tag{1}
$$

The vectors $\mathbf{x}_{lub}$, $\mathbf{x}_{adj}$, $\mathbf{x}_{comp}$ and $\mathbf{x}_{norm}$ denotes the lubrication, adjustment, component and normal operation, respectively. It is possible to occur each of the four events mentioned, through isolated form. The approach consists in designing a classifier for each event.

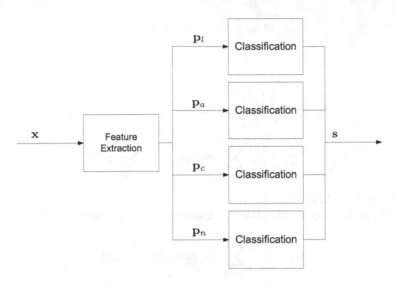

Fig. 1. Block diagram of the scheme for classification of events

2.1   Feature Extraction Based on Higher-Order Statistics (HOS)

Some contributions as [7], related to problems of detection, classification and identification of disturbances in electrical systems, showed significant results obtained through the use HOS. This is because HOS based techniques are better suited to non-Gaussian processes and nonlinear systems, when compared to using second-order statistics.

Considering a sequence $\{z[n]\}$, such that $E\{z[n]\} = 0$. According to [8], the cumulants of the second, third and fourth order can be calculated, respectively, from the following equations (2), (3) and (4).

$$c_{2,z}[i] = E\{z[n]z[n+i]\}, \tag{2}$$

$$c_{3,z}[i] = E\{z[n]z^2[n+i]\}, \tag{3}$$

and

$$c_{4,z}[i] = E\{z[n]z^3[n+i]\} - 3c_{2,z}[i]c_{2,z}[0], \tag{4}$$

where $E\{\cdot\}$ denotes the expected value operator and i is the ith lag.

Assuming that $\{z[n]\}$ is an L-length sequence. Thus, the equations (2) - (4) can be stochastically approximated, respectively, by the following equations (5), (6) and (7).

$$\hat{c}_{2,z}[i] \cong \frac{2}{L} \sum_{n=0}^{L/2-1} z[n]z[n+i], \tag{5}$$

$$\hat{c}_{3,z}[i] \cong \frac{2}{L} \sum_{n=0}^{L/2-1} z[n]z^2[n+i], \tag{6}$$

and

$$\hat{c}_{4,z}[i] \cong \frac{2}{L} \sum_{n=0}^{L/2-1} z[n]z^3[n+i] -$$

$$- \frac{12}{L^2} \sum_{n=0}^{L/2-1} z[n]z[n+i] \sum_{n=0}^{L/2-1} z^2[n], \tag{7}$$

where $i = 0, 1, ..., L/2 - 1$.

An alternative way to calculate the cumulants can be expressed by [9]

$$\tilde{c}_{2,z}[i] \cong \frac{1}{L} \sum_{n=0}^{L-1} z[n]z\,[\mathrm{mod}(n+i, L)] \tag{8}$$

$$\tilde{c}_{3,z}[i] \cong \frac{1}{L} \sum_{n=0}^{L-1} z[n]z^2\,[\mathrm{mod}(n+i, L)] \tag{9}$$

and

$$\tilde{c}_{4,z}[i] \cong \frac{1}{L} \sum_{n=0}^{L-1} z[n]z^3\,[\mathrm{mod}(n+i, L)]$$

$$- \frac{3}{L^2} \sum_{n=0}^{L-1} z[n]z\,[\mathrm{mod}(n+i, L)] \sum_{n=0}^{L-1} z^2[n], \tag{10}$$

where $i = 0, 1, ..., L - 1$ and $\mathrm{mod}(\cdot)$ is the modulus operator.

Thus, for each **s** is obtained a feature vector given by

$$\mathbf{p}_{\mathcal{H}} = [\hat{\mathbf{c}}_{2,z}^T\ \tilde{\mathbf{c}}_{2,z}^T\ \hat{\mathbf{c}}_{3,z}^T\ \tilde{\mathbf{c}}_{3,z}^T\ \hat{\mathbf{c}}_{4,z}^T\ \tilde{\mathbf{c}}_{4,z}^T]^T, \quad \mathcal{H} = 0, 1, \tag{11}$$

where $\mathcal{H} = 0$ is the class without disturbance, while $\mathcal{H} = 1$ is a class with disturbance, $\hat{\mathbf{c}}_{2,z} = [\hat{c}_{2,z}(0), \cdots, \hat{c}_{2,z}(L/2-1)]^T$, $\tilde{\mathbf{c}}_{2,z} = [\tilde{c}_{2,z}(0), \cdots, \tilde{c}_{2,z}(L-1)]^T$, $\hat{\mathbf{c}}_{3,z} = [\hat{c}_{3,z}(0), \cdots, \hat{c}_{3,z}(L/2 - 1)]^T$, $\tilde{\mathbf{c}}_{3,z} = [\tilde{c}_{3,z}(0), \cdots, \tilde{c}_{3,z}(L - 1)]^T$, $\hat{\mathbf{c}}_{4,z} = [\hat{c}_{4,z}(0), \cdots, \hat{c}_{4,z}(L/2 - 1)]^T$ and $\tilde{\mathbf{c}}_{4,z} = [\tilde{c}_{4,z}(0), \cdots, \tilde{c}_{4,z}(L - 1)]^T$.

2.2   Feature Selection Technique Based on Fisher's Discriminant Ratio (FDR)

The feature selection aims to indicate the K_p in order to form the vector $\mathbf{p}_{\mathcal{H}}$, with $\frac{9L}{2}$ length. This makes the complexity of the classifier is quite low. This step is performed only during the design process of classifiers. Recent works, such as [7], have used the Fisher's discriminant ratio (FDR) for feature selection.

In this work we have used the FDR because it is simple and also provides satisfactory results. Its calculation, for a problem involving only two distinct classes is defined in [2] by

$$\mathbf{F}_{FDR} = \Lambda_{\mu_0,\mu_1} \Lambda_\sigma^{-1}, \tag{12}$$

where $\Lambda_\sigma = diag\{\sigma_{0,0}^2 + \sigma_{1,0}^2, \sigma_{0,1}^2 + \sigma_{1,1}^2, \ldots, \sigma_{0,\frac{9L}{2}-1}^2 + \sigma_{1,\frac{9L}{2}-1}^2\}$ is a diagonal matrix composed by the vector covariance associated with each class and $\Lambda_{\mu_0,\mu_1} = diag\{(\mu_{0,0} - \mu_{1,0})^2, (\mu_{0,1} - \mu_{1,1})^2, \ldots, (\mu_{0,\frac{9L}{2}-1} - \mu_{1,\frac{9L}{2}-1})^2\}$ is the diagonal matrix composed by their average vectors.

$\mathbf{v}_{FDR} \in R^{\frac{9L}{2} \times 1}$ is a vector composed by elements from the main diagonal from $\mathbf{F}_{FDR}$, such that $v_{FDR}(0) \geq v_{FDR}(1) \geq \ldots \geq v_{FDR}(\frac{9L}{2} - 1)$, then K_p selected features correspond to K_p first elements of the vector $\mathbf{v}_{FDR}$ are selected as features to constitute the vectors $\mathbf{p}_l$, $\mathbf{p}_a$, $\mathbf{p}_c$ or $\mathbf{p}_n$.

2.3   Classifiers

A simplified description of the used classifiers are presented below.

Bayes Classifier. Consider a vector $\mathbf{x}_p \in \mathbb{R}^{K_p \times 1}$ to be classified among two hypotheses or classes $\mathcal{H}_0$ e $\mathcal{H}_1$, formed by K_p parameters $\mathbf{p}_{\mathcal{H}}$, selected from Equation (12). This vector $\mathbf{x}$ has probability a priori, to be classified into one of two classes, given by $P(\mathcal{H}_0)$ and $P(\mathcal{H}_1)$. The conditional probability density function is denoted by $p(\mathbf{x}|\mathcal{H}_0)$ and $p(\mathbf{x}|\mathcal{H}_1)$. Then, according to [2], Bayes rule provides

$$P(\mathcal{H}_0|\mathbf{x}) = \frac{p(\mathbf{x}|\mathcal{H}_0)P(\mathcal{H}_0)}{p(\mathbf{x})} \tag{13}$$

and

$$P(\mathcal{H}_1|\mathbf{x}) = \frac{p(\mathbf{x}|\mathcal{H}_1)P(\mathcal{H}_1)}{p(\mathbf{x})}, \tag{14}$$

Assume now that the probability a priori of occurs both classes are equal, ie, $P(\mathcal{H}_0) = P(\mathcal{H}_1) = 1/2$ and, the probability density function has a Gaussian distribution. So, one has that:

$$\frac{|\Sigma_0|^{\frac{1}{2}} e^{-\frac{1}{2}(\mathbf{x}-\boldsymbol{\mu}_1)^T \Sigma_1^{-1}(\mathbf{x}-\boldsymbol{\mu}_1)}}{|\Sigma_1|^{\frac{1}{2}} e^{-\frac{1}{2}(\mathbf{x}-\boldsymbol{\mu}_0)^T \Sigma_0^{-1}(\mathbf{x}-\boldsymbol{\mu}_0)}} \underset{<}{\overset{\geq}{}} 1 \tag{15}$$

It is observed in (15), that the proposed classifier is based on the ML criterion and, given a vector $\mathbf{x}$, the determination of which class it corresponds depends on the result of this inequality.

Type-1 and Singleton Fuzzy Logic System. Assuming the choice for singleton fuzzification, max-product composition, product implication and height defuzzifier and leaving open the choice of membership function, is simple to show that the output of the type-1 and singleton Fuzzy Logic System [5] is defined by equation (16)

$$y(\mathbf{x}) = f_s(\mathbf{x})$$
$$= \frac{\sum_{l=1}^{M} \theta_l \prod_{k=1}^{K_p} \mu_{F_k^l}(x_k)}{\sum_{l=1}^{M} \prod_{k=1}^{K_p} \mu_{F_k^l}(x_k)}, \tag{16}$$

where $\mathbf{x} \in \mathbb{R}$ is the vector constituted by the features extracted from the vector $\mathbf{x}$ and θ_l is the weight associated with l-th rule, $l = 1, \ldots, M$. Note that the use of subscript "s" on $f_s(\mathbf{x})$ remits that this is a type-1 and singleton FLS. Knowing that t-norms is a product operator and each $\mu_{F_k^l}(x_k)$ can be assumed as a Gaussian membership function, then

$$\mu_{F_k^l}(x_k) = \exp\left\{ -\frac{1}{2}\left(\frac{x_k - m_{F_k^l}}{\sigma_{F_k^l}} \right)^2 \right\}, \tag{17}$$

where $m_{F_k^l}$ and $\sigma_{F_k^l}^2$ denotes the mean and variance, respectively.

For designed FLS, it was decided to make use of the training algorithm called Backpropagation. Given a set of input-output pairs $(\mathbf{x}^{(q)} : y^{(q)})$, it is aimed to establish a solution that leads to an optimal setting for such sets, with the backing of the cost function, which is expressed by [5]

$$J(\mathbf{w}^{(q)}) = \frac{1}{2}\left[f_s(\mathbf{x}^{(q)}) - y^{(q)} \right]^2, \tag{18}$$

is minimized. The minimization of the cost function results in

$$m_{F_k^l}(q+1) = m_{F_k^l}(q) - \alpha_m \left[f_s(\mathbf{x}^{(q)}) - y^{(q)} \right] \times$$
$$\left[\theta_l(q) - f_s(\mathbf{x}^{(q)}) \right] \frac{\left[x_k^{(q)} - m_{F_k^l}(q) \right]}{\sigma_{F_k^l}^2(q)} \phi_l(\mathbf{x}^{(q)}), \tag{19}$$

$$\theta_l(q+1) = \theta_l(q) - \alpha_{\theta_l(q)} \left[f_s(\mathbf{x}^{(q)}) - y^{(q)} \right] \phi_l(\mathbf{x}^{(q)}), \tag{20}$$

and

$$\sigma_{F_k^l}(q+1) = \sigma_{F_k^l}(q) - \alpha_\sigma \left[f_s(\mathbf{x}^{(q)}) - y^{(q)} \right] \times$$
$$\left[\theta_l(q) - f_s(\mathbf{x}^{(q)}) \right] \frac{\left[x_k^{(q)} - m_{F_k^l}(q) \right]^2}{\sigma_{F_k^l}^3(q)} \phi_l(\mathbf{x}^{(q)}). \tag{21}$$

MLP Neural Network Classifier. Given the vectors $\mathbf{x}_{p,i}$, with $i = 1, 2, \ldots, N_x$ samples, constituted by K_p extracted features from vectors $\mathbf{p}_{x,i}$, then the equations of a MLP neural network with K_p inputs, one hidden layer and one output layer, are the following:

$$\mathbf{s}_i = \mathbf{A}^T \begin{bmatrix} \mathbf{x}_{p,i} \\ 1 \end{bmatrix} \tag{22}$$

$$\mathbf{q}_i = \varphi(\mathbf{s}_i), \tag{23}$$

and

$$y_{rn,i} = \mathbf{B}^T \begin{bmatrix} \mathbf{q}_i \\ 1 \end{bmatrix}, \tag{24}$$

where $\mathbf{A} \in \mathbb{R}^{(K_p+1) \times N_q}$ and $\mathbf{B} \in \mathbb{R}^{(K_p+1) \times 1}$ are the weights matrix between the input and hidden/ intermediate layers and between the hidden/intermediate and output layers, respectively; $y_{rn,i}$ is the output of the MLP neural network associated to the vector $\mathbf{x}_{p,i}$.

In turn, $\varphi(\cdot)$ is a monotonically increasing transfer function and that this work will be taken as the hyperbolic tangent. Thus, Equation (23) can be rewritten as follows:

$$\mathbf{q}_i = \tanh(\mathbf{s}_i). \tag{25}$$

Concatenating the column vectors of the matrices $\mathbf{A}$ and $\mathbf{B}$ is obtained the weight vector given by

$$\mathbf{w} = [\mathbf{a}^T \ \mathbf{b}^T]^T, \tag{26}$$

where $\mathbf{a}^T$ and $\mathbf{b}^T$ is the concatenation of column vectors of the matrices $\mathbf{A}$ and $\mathbf{B}$, respectively. The optimal vector $\mathbf{w}$, given by $(\mathbf{w}_o)$, can be obtained by

$$\mathbf{w}_o = \min_{\mathbf{w}} J(\mathbf{w}), \tag{27}$$

where

$$J(\mathbf{w}) = \frac{1}{2N_x} \sum_{i=1}^{N_x} (y_{rn}(i) - y_d(i))^2 \tag{28}$$

is the cost function to be minimized and $y_d(i)$ is the ith desired output of the MLP neural network. Note that $y_d(i) = 1$ means that the parameter vector $\mathbf{x}_{p,i}$ is associated with the occurrence of the event. On the other hand, $y_d(i) = -1$ states that the parameter vector $\mathbf{x}_{p,i}$ is associated with the absence of event.

Among several training methods available in the literature for obtaining $\mathbf{w}_o$, in this work we opted for Levenberg-Marquardt (LM) [10]. A detailed description of the LM algortith for training the MLP Neural Network is presented in [11].

3   Experimental Results

The data provided by MRS Logística S.A. are coming from the signature of SM.

The current [A] of Switch Machine was acquired through industrial data collector, with sampling rate of 100 Hz for a period of 2 seconds, totaling 200 samples. This time is sufficient for the equipment perform their duties. Figure 2 exemplifies samples acquired from the signal. Despite this list of combinations of events not be complete, note that it is representative to illustrate the complexity of the problem of classification of events and provide an excellent interpretation of the trouble by the maintenance team.

The acquired signals refer to the following types of classes: Defect due to lack of lubrication, lack of adjustment, defect in some mechanical component and normal operation of the SM. Originally, the database containing the signals to be classified were unbalanced. This occurrence is due to the greater incidence of normal operation, since the SM is in a satisfactory state of operation. The original data from lubrication, adjustment and component were balanced aimed at numerical equality with the database constituted by signals of normal operation. Additionally, were removed outliers from incorrect analysis (erroneous values) taken by specialist of MRS Logística S.A., when the database was generated. Upon completion, thus, is available 1376 signals for each of the four classes proposed. Such signals can be viewed in Figure 2.

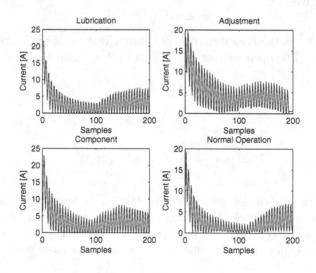

Fig. 2. Signals referring to used classes

The parameters of the type-1 and singleton FLS are as follows: $M = 4$ (two rules for class that has the presence of the event and two rules for the class that does not have the presence of the event); $K_p = 3$, since the FDR was used to select the feature vector of lower dimensionality possible for the classification problem. The initialization of the parameters from membership functions was previously defined heuristically from the calculation of means and variances of the coefficients of feature vectors.

The MLP Neural Network has $K_p = 3$ inputs, $N_q = 1$ hidden layer with two processors and 1 output layer with one processor.

Additionally, for the simulations of the type-1 and singleton FLS and MLP Neural Network, it was considered 100 epochs for training algorithm and applied cross-validation. For type-1 and singleton FLS, in order to obtain the results to be presented, it was decided to divide the data equally for obtaining the training and test sets. Since we have 400 samples for each of the three parameters extracted, is obtained therefrom 200 samples for the training set and 200 samples for the test set. For the MLP Neural Network, it was opted for the use of a validation set and the division of the data was taken as follows: from a total of 400 samples, note that 200 samples were used for training, 100 for validation and 100 for testing. Also, it was opted for check the error for 6 epochs of training, in order to implement early stopping if the error increase. It is important to worth that other simulations were performed in order to ensure that the results obtained can be reached again.

3.1   Analysis of Performance and Convergence

The results obtained for the classifiers based on type-1 and singleton FLS and MLP Neural Network are shown in Tables 1 and 2, respectively. The final results from comparison between the techniques are shown in Table 3, which is portrayed the performance during the test of the proposed techniques. The efficiency ρ of each classifier is the product of the performances obtained for all events.

Table 1. Performance in (%) of the classifier based on the type-1 and singleton FLS

Fuzzy		
Events	Training	Test
Lubrication	100,0	100,0
Adjustment	97,5	97,5
Component	98,5	98,5
Normal Operation	99,5	99,5

Table 2. Performance in (%) of the classifier based on MLP Neural Network

MLP Neural Network			
Events	Training	Validation	Test
Lubrication	99,6	100,0	100,0
Adjustment	100,0	100,0	100,0
Component	99,6	98,3	98,3
Normal Operation	100,0	100,0	100,0

Table 3. Comparative performance in (%) of the classifiers adopted during the test phase

Comparison Between Proposed Methods			
Events	Bayes	Fuzzy	MLP
Lubrication	94,5	100,0	100,0
Adjustment	98,0	97,5	100,0
Component	99,0	98,5	98,3
Normal Operation	98,5	99,5	100,0
Efficiency (ρ)	97,5	98,9	99,6

4   Conclusions

This paper discussed the use of Pattern Recognition and Computational Intelligence (CI) techniques to classify defects in SM. The extracted features by EOS and selected features by FDR are relevant and have contributed to a reduction of dimensionality of the data to be presented for the proposed classifiers.

The computational results obtained from real data show that the MLP Neural Network has better convergence and performance compared to the Bayes classifier and type-1 and singleton FLS. It is noteworthy that the performance of Bayes classifier is slightly inferior to the other classifiers. However, the said classifier proves to be an attractive option due to the mathematical simplicity, even having the lowest efficiency among the presented proposals.

The next activities are aimed at creating and implementing a real-time application, to be integrated into the existing SM supervision system in the company MRS Logística S.A..

Acknowledgments. The authors would like to thank MRS Logística S.A. for providing financial support.

References

1. Duda, R.O., Hart, P.E., Stork, D.G.: Pattern Classification. John Wiley & Sons, New York (2001)
2. Theodoridis, S., Koutroumbas, K.: Pattern Recognition. Academic Press, San Diego (1999)
3. Shao, W., Bouzerdoum, A., Phung, S.L., Su, L., Indraratna, B., Rujikiatkamjorn, C.: Automatic Classification of Ground-Penetrating-Radar Signals for Railway-Ballast Assessment. IEEE Trans. on Geoscience and Remote Sensing 49(10), 3961–3972 (2011)
4. Feng, H., Jiang, Z., Xie, F., Yang, P., Shi, J., Chen, L.: Automatic Fastener Classification and Defect Detection in Vision-Based Railway Inspection Systems. IEEE Trans. on Instrumentation and Measurement 63(4), 877–888 (2014)
5. Mendel, J.M.: Uncertain Rule-Based Fuzzy Logic Systems: Introduction and New Directions. Prentice Hall PTR (2001)

6. Haykin, S.: Neural Networks. A Comprehensive Foundation. Prentice Hall (1999)
7. de Aguiar, E.P., Marques, C.A.G., Duque, C.A., Ribeiro, M.V.: Signal decomposition with reduced complexity for classification of isolated and multiple disturbances in electric signals. IEEE Trans. on Power Delivery 24(4), 2459–2460 (2009)
8. Mendel, J.M.: Tutorial on higher-order statistics (spectra) in signal processing and system theory: theoretical results and some applications. Proceedings of the IEEE 79(3), 278–305 (1991)
9. Ribeiro, M.V., Marques, C.G., Cerqueira, A.S., Duque, C.A., Pereira, J.L.R.: Detection of Disturbances in Voltage Signals for Power Quality Analysis Using HOS. EURASIP Journal on Advances in Signal Processing, 13–15 (2007)
10. Hagan, M.T., Menhaj, M.-B.: Training feedforward networks with the Marquardt algorithm. IEEE Trans. on Neural Networks. 5(6), 989–993 (1994)
11. Finschi, L.: An Implementation of the Levenberg-Marquardt Algorithm. ETH Zurich - Institute for Operations Research (1994)

An Accurate Flood Forecasting Model Using Wireless Sensor Networks and Chaos Theory: A Case Study with Real WSN Deployment in Brazil

Gustavo Furquim[1], Rodrigo Mello[1], Gustavo Pessin[2], Bruno S. Faiçal[1], Eduardo M. Mendiondo[3], and Jó Ueyama[1]

[1] Institute of Mathematics and Computer Science (ICMC)
University of São Paulo (USP) - São Carlos, SP, Brazil
{gafurquim,mello,bsfaical,joueyama}@icmc.usp.br
[2] Vale Institute of Technology
Belém, PA, Brazil
gustavo.pessin@itv.org
[3] Sao Carlos School of Engineering (EESC)
University of Sao Paulo (USP).
University of São Paulo (USP) - São Carlos, SP, Brazil
emm@sc.usp.br

Abstract. Monitoring natural environments is a challenging task on account of their hostile features. The use of wireless sensor networks (WSN) for data collection is a viable method since these domains lack any infrastructure. Further studies are required to handle the data collected to provide a better modeling of behavior and make it possible to forecast impending disasters. These factors have led to this paper which conducts an analysis of the use of data gathered from urban rivers to forecast future flooding with a view to reducing the damage they cause. The data were collected by means of a WSN in São Carlos, São Paulo State, Brazil and were handled by employing the Immersion Theorem. The WSN were deployed by our group in the city of São Carlos due to numerous problems with floods. After discovering the data interdependence, artificial neural networks were employed to establish more accurate forecasting models.

Keywords: Wireless Sensor Network, Machine Learning, Time series Analysis, Chaos Theory, Modeling, Prediction.

1 Introduction

Natural disasters, such as landslides, serious floods, fires, volcanic eruptions and the damage they cause, are global problems, which incur a heavy cost in terms of human lives and financial losses. Every year about 102 million people around the world are affected by the problem of flooding and it is expected that this number will increase in the years ahead. The regions that suffer most from floods, are developing countries and urban areas [1]. These are the characteristics of

V. Mladenov et al. (Eds.): EANN 2014, CCIS 459, pp. 92–102, 2014.

São Carlos in Brazil where the local climate has been altered by pollution, which has destroyed the eco-system.

Wireless sensor networks are a beneficial means of carrying out the monitoring of urban rivers and other natural environments since they have a number of attractive features: low cost, particularly with regard to infrastructure, low energy consumption, the provision of access to inhospitable surroundings, the fact that they are simple to install and high-precision sensors are employed which are adaptable to environmental change [2], [3]. A WSN called REDE has been constructed and deployed by our group in the town of São Carlos. This was developed by the Institute of Mathematical Sciences and Computing, University of Sao Paulo (USP) and integrates the e-NOE project (WSN for monitoring urban rivers) which aims to carry out the monitoring of urban rivers [3], [4]. To date, our WSN system is only capable of detecting floods when it has already taken place. Hence, we are now keen to enable it to predict floods before it has taken place, which is one of the aims of this paper. By doing so, we can ensure that the population at risk are evacuated before the floods. This measure can help us to reduce the problems arising from floods.

The nodes of the REDE system provide measurements of the water pressure at the bottom of the river by converting this value into centimeters to describe height. These values are collected at regular time intervals and can thus describe the height of the river over a period of time. This analysis of a temporal series makes it possible to study, model and predict systems with these features. On the basis of this analysis, the time series can be defined as data that is collated in terms of observable variables over a period of time [5], [6] or in other words, the time series is an orderly sequence of observations collected at regular intervals. Thus the data collected by the REDE system constitutes a time series which, therefore, can be studied in the light of the concepts from Time Series Analysis. This paper is based on the assumption that there is a temporal relationship among the observations of the level of the river. This means that the value of an observation at a current instant, depends to some extent on past values. This paper seeks to determine how to find out this temporal relationship, as well as to model and forecast the level of the rivers. In achieving this end, it employs tools that originate from the Time Series Analysis or more precisely, the sub-area of Chaos Theory. In this research, we employed Chaos Theory to help in the unfolding of the time series. The unfolding techniques are employed to find different behavior which can not be observed while viewing it in fewer dimensions. In this sense, the problem of flooding might not be chaotic itself although the chaos theory helps in the understanding of the time series behavior.

One of the main results from this sub-area is the Immersion Theorem put forward by Takens [7] which allows the temporal series to be reconstructed in vectors with m values (m is the designated embedded dimension). Each of these values corresponds to an observation spaced out in intervals in accordance with a time delay or separation dimension called τ. As several studies have shown, these vectors represent the interdependent relations among observations that increase the accuracy of the modeling and hence the prediction of the time series.

For example, when considering a time series $X = (x_{-\beta}, ..., x_{-1}, x_0)$ observed in the period of time $[-\beta, 0]$, where $-\beta < 0$, each vector that is reconstructed by the Immersion Theorem corresponds to $I = (x_{-\tau \times (m-1)}, ..., x_{2 \times \tau}, x_0)$. After obtaining those vectors, modeling techniques can be employed that allow the observations to be related over a period of time and thus forecasting to be carried out. Hence, before obtaining these vectors, it is necessary to estimate values for the embedded dimension (m) and for separation dimension (τ). Further details are given in Section 3. After the reconstruction of the time series has been undertaken by means of Takens' Immersion Theorem [7], machine learning techniques such as Artificial Neural Networks (ANN), Support Vector Machines (SVM), or even a combination of several techniques can be used to model the data and thus lead to forecasting at a higher degree of accuracy. The Multilayer Perceptron (MLP) [8] technique from the artificial neural network was employed in this paper to carry out the forecasting, owing to the good results obtained in the predictions when this was used together with the data gathered by the WSNs in the urban rivers [9].

The remainder of this paper is divided as follows: Section 2 examines some related studies that employ the Chaos Theory, WSNs or include flood forecasting. In Section 3, the tools of the Chaos Theory and the Immersion Theorem are explained in greater detail. Section 4 outlines the proposed method and describes the WSN that is used, the handling of the data and the artificial neural network used in this study. This work ends with an analysis of the results obtained (Section 5) followed by the conclusion and considerations on future work (Section 6).

2   Related Works

Seal [2] presents a flood forecast scheme using a hybrid approach (centralized and distributed) for WSNs to rivers. The WSN architecture consists of several sensors collecting data combined with processor nodes, where the forecast algorithm is implemented, and centers of manually operated monitoring, which implement the redundancy by comparing the real situation and the forecast and initiates the evacuation procedures. The data flow is, basically: data collection, calculation of the coefficient for regression, coefficient update for regression, sending the results and then informing community. The forecast model uses robust linear regression, being independent of the number of parameters. Besides the results, Seal [2] shows other related studies, making clear the applicability of WSNs to collect data for flood forecasting.

In [10] an architeture to monitor and predict slope disaster is proposed. This architeture is based on WSNs and mobile communication to transmit warnings, indicating areas with the possibility of disasters. The main elements of the architeture are: (i) Mobile user site, implementing the user interface, (2) Hillslopes monitoring sensor site, the WSN physical implementation, (3) Integrated service server, providing services like network integration and (4) Intelligent hillslopes decision system, which predict the hillslopes degree of hazard. The model user

Analytic Network Process (ANP) to predict the danger degree and has an accuracy of 88,33%.

Ishii [5] proposes an online prediction approach to support data access optimization on distributed systems. The data acquisition about processes was built using the Unix DLSym library and transformed into multidimensional time series. Time series were then analyzed to evaluate the best model to make predictions. As result, Chaos Theory tools were used to unfold data and the Radial Basis Function (RBF) neural network was used to model time series.

3   Chaos Theory Concepts

Takens [7] observed that a time series $x_0, x_1, ..., x_{n-1}$ can be reconstructed in a multidimensional space $x_n(m, \tau) = (x_n, x_{n+\tau}, ..., x_{n+(m-1)\tau})$, also called time-delay coordinate space, where m is the embedded dimension and τ represents the time delay (or separation dimension). This mapping or reconstruction technique allows to transform dynamical system observations (or rule outputs) in a set of points in an m-dimensional Euclidean space. This reconstruction supports the obtainment of dynamical systems rules, consequently simplifying the study of behaviors and their usage under different circumstances, such as the study of orbits, tendencies and prediction [11].

To better understand the embedded (number of dimensions) and separation (time delay) dimensions, consider the Logistic map outputs, previously approached, reconstructed in a multidimensional space where $m = 2$ and $\tau = 1$, which results in pairs of points (x_t, x_{t+1}) (Figure 1 (b)). After the reconstruction, the Logistic map behavior, which behaved as random walk (Figure 1 (a)), can be studied, understood and modeled in a simpler way. After making a data regression, we can obtain the dynamic system rule and, therefore, we are able to understand transitions, estimate and predict observations. Having such rule and an initial x_t, we can, for example, define the next series observation, x_{t+1}, which feeds the approach back and generates x_{t+2}, and, thus, consecutively.

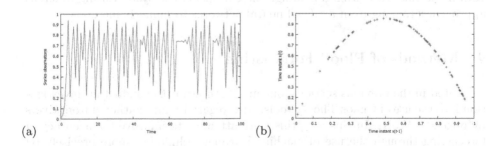

(a)   (b)

Fig. 1. (a) Logistic map outputs – First 100 observations and (b) observations of the reconstructed Logistic map (embedded dimension 2 and separation 1)

The embedded dimension basically defines the number of axis for the time-delay coordinate space. This determines the number of dimensions necessary to unfold the reconstructed series. In this case, the series required two dimensions, others may need more. This behavior is, for example, observed in the Lorenz attractor which requires three dimensions [12]. Besides the embedded dimension, there is still the separation one, which supports the extraction of the periodicity of series behavior. This dimension informs the time delay of historical observations to be modeled and analyzed in order to predict future events (it basically allows to points out how far we should go to obtain cause-consequence relationships in the series). The embedded and separation dimensions support the study of series, however, we need to find those dimensions for any series, including the ones generated by experimental data. According to Fraser and Swinney [13], the Auto-Mutual Information technique (AMI) presents better results when estimating the separation dimension. To obtain the series separation, one employs AMI under different time delays. Afterwards, one plots a curve in function of time delays (starting at 1 and incrementing) and considers the first minimum as main candidate for the separation dimension [14]. Besides that is a good candidate, one should also consider other separation dimension values and observe the characteristics of the time-delay coordinate space obtained.

After defining the separation dimension, we must find the the embedded one. Kennel [15] proposed the False Nearest Neighbors method (FNN) to estimate the embedded dimension. This method computes the closest neighbors for every data point in the time-delay coordinate space, starting with embedded dimension equals to 1. Afterwards, a new dimension is added and the distance among the closest neighbors is again calculated. When this distance increases, data points are considered False Nearest Neighbors, what makes evident the need of more dimensions to unfold the series behavior [14]. After defining both dimensions, we employ the Immersion Theorem by Takens [7], as previously presented, where the time series $x_0, x_1, ..., x_{n-1}$ is reconstructed in a multidimensional space, or time-delay coordinate space, $x_n(m, \tau) = (x_n, x_{n+\tau}, ..., x_{n+(m-1)\tau})$ (the component m represents the embedded dimension, this is, the number of dimensions to unfold the series, and τ is the separation, this is, the time delay to consider historical observations). The reconstruction unfolds the dynamical system, what allows to obtain the rule. After this unfolding, one can model the system using different approaches, ranging from artificial neural networks to numeric methods.

4 Methods of Flood Forecasting

As stated in the previous sections, our main concern is to conduct an investigation into the use of Chaos Theory tools, with regard to data gathered from rivers by means of WSNs. In this way, this work attempts to improve the accuracy of forecasting through the use of machine learning techniques, more precisely by artificial neural networks. The WSN installed in the town of São Carlos, Brazil comprises: (i) sensors 1, 2 and 3 which provide measurements of the pressure at the bottom of the river and converts this value into the height as centimeters

(ii) a router capable of increasing the communication distance and allowing the sensors to be positioned in a wider area and (iii) a base station that picks up the information from several sensors and allows a better analysis of the data to be conducted. As well as measuring the height of the river level, the sensor 3 is equipped with a photographic camera which makes visual information available about the river conditions. In this paper, data was collected by Sensor 1 during a period of one month (from 10/01/2013 to 10/31/2013), at intervals of 5 minutes between the measurements. The behavior of the river level is illustrated in Figure 2, where time is described in a day:hour:minute format. The required outcome (following the data handling, and modeling of the system through the Multilayer Perceptron artificial neural network) is the level of the river in the next instants.

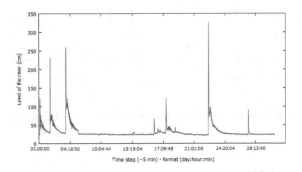

Fig. 2. Level of the river during March 2013

Adopting this approach, the first stage is to calculate the Auto-Mutual Information technique (AMI) to determine the value of the separation dimension (time delay τ). As proposed in [16], one should select the first minimum of this graph from left to right as the separation dimension, which is $\tau = 21$. However, as observed in other studies [15], [14], when AMI slightly reduces as the separation dimension increases, a good attempt for the separation dimension would be $\tau = 1$. In this work, we consider both values for τ in order to provide different immersions and verify the best. After computing the separation dimension, the next stage is to estimate the embedded dimension (m) through the False Nearest Neighbors (FNN) method [15]. According to this method, the embedded dimension is selected when fraction of false neighbors is equal to zero, however this is very hard to obtain for real-world data, as it may contain noise. In noisy scenarios, we can select the number of dimensions in which the fraction of false neighbors becomes less than 30% as presented in [17]. Figure 3 shows the results of FNN when considering $\tau = 1$ (a) and $\tau = 21$ (b), in which the x-axis corresponds to embedded dimensions and the y-axis is related to the fraction of false neighbors. Thus, we selected $m=4$ when $\tau = 1$ and $m=11$ when $\tau = 21$. Greater values can be considered for the embedded dimension, however, according to Kennel [15], this will not have much influence because once the behavior of the time series has unfolded, it can be understood and studied. After the reconstruction of the series, the Multilayer Perceptron implemented in WEKA [18] was employed to model it. We set parameters as learning rate = 0.3, 2 hidden layers with 3 neurons each when $m=4$ and $\tau = 1$ and 5 hidden layers with 10 neurons each when $m=11$ when $\tau = 21$ and performed tests using cross-validation. Parameters employed to build the ANN were the default values obtained from WEKA. Section 5 presents the experimental results.

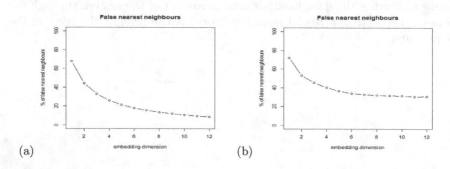

(a)   (b)

Fig. 3. Percentage of false neighbors for the river level, considering (a) $\tau = 1$ and (b) $\tau = 21$

5   Results and Discussion

Table 1 shows the mean squared error and the coefficient of determination (R^2) obtained when $\tau = 1$ with $m = 4$ and $\tau = 21$ with $m = 11$. The mean squared error and the coefficient of determination were used because they help to determine the model quality. It can be confirmed that the mean squared error obtained by means of $\tau = 21$ and $m = 11$ is more than 6 times greater than the mean squared error with $\tau = 1$ and $m = 4$ as parameters for the unfolding. The coefficient of determination also confirms the better results for $\tau = 1$ and $m = 4$. Through these experiments, it can be observed that, since the datasets have been obtained from a real environment that is subject to noise, the Auto-Mutual Information only estimates what are good values for the separation dimension. However, other minimum points and the value of $\tau = 1$ must be regarded as options.

Figures 4 and 5 show the behavioral pattern of the river during the month of March (red line) and the values predicted by MLP when trained with the data (green line). It can be seen that in Figure 4 where the values for $\tau = 1$ and $m = 4$ are shown, for the most part, the lines are overlapping, which suggest that the predicted values are closer than expected. In Figure 5, in which $\tau = 11$ with $m = 21$ are employed, the lines end up being divergent, particularly at the peaks which are important regions because they show instants where there is a possibility of flooding.

Table 1. Multilayer Perceptron: mean squared error

	MLP mean squared error (cm^2)	Coefficient of determination
$m = 1$ and $\tau = 4$	1.5732	0.9618
$m = 11$ and $\tau = 21$	9.4761	0.1687

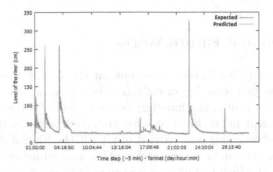

Fig. 4. Expected river level and predicted river level for $\tau = 1$ and $m = 4$

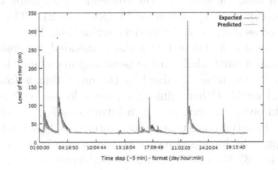

Fig. 5. Expected river level and predicted river level for $\tau = 11$ and $m = 21$

Considering the best results achieved with $\tau = 1$ and $m = 4$, we applied a recursive prediction using this parametrization. These experiments used values previously predicted to feed the model and estimate the next six time instants, i.e., river levels. For example, considering x_0, x_1 and x_2 the measured river levels at the instants t_0, t_1 and t_2, where $t_0 < t_1 < t_2$, and p_3 the value predicted by the MLP for the next instant t_3. We used x_1, x_2 and p_3 to predict the value of the river level at the instant t_4, in other words, p_4. Then, x_2, p_3 and p_4 were used in the next prediction and so on. Table 2 shows the mean squared error and the coefficient of determination (R^2) obtained for each step of recursive prediction.

Table 2. Recursive Prediction using Multilayer Perceptron: mean squared error

Prediction	Second	Third	Fourth	Fifth	Sixth
MLP mean squared error (cm^2)	24.6471	28.326	29.669	29.4479	32.1645
Coefficient of determination	0.3174	0.2198	0.1633	0.137	0.0995

6   Conclusion and Future Works

In this paper, a study has been carried out on the use of Takens' Immersion Theorem [7] in the preprocessing of collected data from urban rivers by means of WSNs and subsequently, the handling of these data for flood prediction by using artificial neural networks. The data collected were analyzed as a time series and the preprocessing was carried out with the aim of determining the interdependence of the data in the series. This procedure enabled a simpler model to be devised which had greater accuracy in forecasting.

Two values for the separation dimension ($\tau = 1$ and $\tau = 21$) were evaluated and on the basis of this, the values for the embedded dimension were calculated as follows: $m = 4$ (for $\tau = 1$) and $m = 11$ (for $\tau = 21$). The mean squared error value was used as a criterion for models evaluation. The use of values from the separation dimension as being 1 and the embedded dimension as being 4, obtained better results particularly in the peak regions where there is a greater risk of floods. This fact can be explained by the noise that is found in the data gathered from real-world. Other minimum points found by the Auto-Mutual Information should be taken into account in future studies, as well as techniques that are able to reduce the noise that is encountered in this kind of time series. As expected, despite the increased mean squared error in recursive prediction, we can consider that 32.1645 cm^2 (or 5.6714 cm) error in the sixth prediction (approximately 30 minutes after the last reading) indicates that good results can be achieved using the model created in a recursively manner.

In future work we intend to broaden our examination of the machine learning techniques that can be used on the preprocessed data to model the system. In addition, readings from other sensors in the REDE system (Sensor 2 and 3) were not used and might contain information that could improve the accuracy of the forecasting. Moreover, a rain gauge has been incorporated in the REDE system and is currently undergoing tests which will allow it to collect data about the rainfall in the region. A study that applies the Immersion Theorem for a set of data formed of both the readings of the sensors and the readings of the rain gauge would be of great interest and could lead to more precise models being devised. In future, we also seek to embed this prediction model in the sensors and thus make it possible to take action in a more independent way in extreme situations, such as a breakdown in communication or the destruction of nodes. In making this integration, the low consumption of energy of each node should be taken into account and is a factor requiring further study.

Acknowledgments. The authors would like to acknowledge the financial support granted by FAPESP, process ID 2012/22550-0 and RNP (National Research Network) - CIA2-RIO.

References

1. de Freitas, C.M., Ximenes, E.F.: Floods and public health – a review of the recent scientific literature on the causes, consequences and responses to prevention and mitigation. Ciência e Saúde Coletiva 17, 1601–1616 (2012)
2. Seal, V., Raha, A., Maity, S., Mitra, S.K., Mukherjee, A., Naskar, M.K.: A simple flood forecasting scheme using wireless sensor networks. CoRR abs/1203.2511 (2012)
3. Ueyama, J., Hughes, D., Man, K.L., Guan, S., Matthys, N., Horre, W., Michiels, S., Huygens, C., Joosen, W.: Applying a multi-paradigm approach to implementing wireless sensor network based river monitoring. In: 2010 First ACIS International Symposium on Cryptography and Network Security, Data Mining and Knowledge Discovery, E-Commerce &#38 Its Applications and Embedded Systems (CDEE), pp. 187–191 (October 2010)
4. Hughes, D., Ueyama, J., Mendiondo, E., Matthys, N., Horré, W., Michiels, S., Huygens, C., Joosen, W., Man, K., Guan, S.-U.: A middleware platform to support river monitoring using wireless sensor networks. Journal of the Brazilian Computer Society 17(2), 85–102 (2011)
5. Ishii, R.P., de Mello, R.F.: An online data access prediction and optimization approach for distributed systems. IEEE Transactions on Parallel and Distributed Systems 23(6), 1017–1029 (2012)
6. Mello, R.: Improving the performance and accuracy of time series modeling based on autonomic computing systems. Journal of Ambient Intelligence and Humanized Computing 2(1), 11–33 (2011)
7. Takens, F.: Detecting strange attractors in turbulence. In: Rand, D., Young, L.-S. (eds.) Dynamical Systems and Turbulence, Warwick 1980. Lecture Notes in Mathematics, vol. 898, pp. 366–381. Springer, Heidelberg (1981)
8. Haykin, S.: Neural Networks: A Comprehensive Foundation, 2nd edn. Prentice Hall PTR, Upper Saddle River (1998)
9. Furquim, G., Neto, F., Pessin, G., Ueyama, J., Clara, M., Mendiondo, E.M., Souza, P., Dimitrova, D., Braun, T.: Combining wireless sensor networks and machine learning for flash flood nowcasting. Int. Workshop on Bio and Intelligent Computing (2014)
10. C.-I. Wu, H.-Y. Kung, C.-H. Chen, and L.-C. Kuo, "An intelligent slope disaster prediction and monitoring system based on wsn and anp," *Expert Systems with Applications*, 2014.
11. Alligood, K., Sauer, T., Yorke, J.: Chaos: An Introduction to Dynamical Systems. New York, NY (1997)
12. Lorenz, E.N.: Deterministic Nonperiodic Flow.. Journal of Atmospheric Sciences 20, 130–148 (1963)
13. Fraser, A.M., Swinney, H.L.: Independent coordinates for strange attractors from mutual information. Physical Review A 33, 1134–1140 (1986)
14. Mello, R., Yang, L.: Prediction of dynamical, nonlinear, and unstable process behavior. The Journal of Supercomputing 49(1), 22–41 (2009)

15. Kennel, M.B., Brown, R., Abarbanel, H.D.I.: Determining embedding dimension for phase-space reconstruction using a geometrical construction. Phys. Rev. A 45, 3403–3411 (1992)
16. Abarbanel, H.D.I., Brown, R., Sidorowich, J.J., Tsimring, L.S.: The analysis of observed chaotic data in physical systems. Rev. Mod. Phys. 65 (1993)
17. Liebert, W., Pawelzik, K., Schuster, H.G.: Optimal embeddings of chaotic attractors from topological considerations. Europhysics Letters 14 (1991)
18. Hall, M., Frank, E., Holmes, G., Pfahringer, B., Reutemann, P., Witten, I.H.: The weka data mining software: an update. SIGKDD Explor. Newsl. (2009)

Regenerative Braking Control Strategy for Hybrid and Electric Vehicles Using Artificial Neural Networks

Sanketh S. Shetty[1] and Orkun Karabasoglu[2,3,*]

[1] Department of Electrical Engineering, IIT Roorkee, Roorkee,
247 667 Uttarakhand, India
[2] Department of Electrical and Computer Engineering
Sun Yat-Sen University - Carnegie Mellon University
(SYSU-CMU) Joint Institute of Engineering, Pittsburgh, PA, USA
[3] SYSU-CMU Shunde International Joint Research Institute, Guangdong, China
sankyuee@iitr.ernet.in, karabasoglu@cmu.edu

Abstract. One of the fundamental advantages of hybrid and electric vehicles compared to conventional vehicles is the regenerative braking mechanism. Some portion of the kinetic energy of the vehicle can be recovered during regenerative braking by using the electric drive system as a generator with the appropriate control strategy. The control requires distribution of the brake forces between front and rear axles of the vehicle and also between regenerative braking and frictional braking. In this paper, we propose solving the optimal brake force distribution problem using an Artificial Neural Network based methodology in order to maximize the available energy for recovery while following the rules for stability. Using the proposed approach, we find that for urban driving pattern, UDDS, up to 37 % of the total energy demand can be recovered. Then we compare the amount of recovered energy for different driving cycles and show that aggressive driving reduces recoverable energy up to 7%. An increase in the energy recovery rate directly translates into improvements in fuel economy and reductions in emissions.

Keywords: Regenerative braking, artificial neural networks, brake force distribution, electric vehicle.

1 Introduction

In a conventional, internal combustion (IC) engine powered vehicle, the power flow occurs from the IC engine towards the wheels via a transmission system. The mechanical energy that is produced is used up to overcome aerodynamic drag, rolling resistance, frictional forces and to accelerate the vehicle [1]. These are all unavoidable components of energy expenditure. In the case of braking or deceleration of the vehicle, the kinetic energy of the vehicle that was built up during acceleration phase has to be reduced in order to reduce the speed of the vehicle. However since there is no mechanism in place to convert this kinetic energy into some other form to store away, all of the energy gets dissipated in the form of heat via the friction brakes. This

---

* Corresponding author.

V. Mladenov et al. (Eds.): EANN 2014, CCIS 459, pp. 103–112, 2014.
© Springer International Publishing Switzerland 2014

energy is lost and additional fuel needs to be burnt when the vehicle needs to accelerate again. In urban environments, where vehicles are subjected to intermittent start-stop motion, this effect is even worse due to the higher frequency of application of brakes [2]. The problem is the inability of the conventional vehicle to recover energy while braking. This ultimately leads to low fuel economy and has negative economic implications especially given the size of transportation sector.

Electrified vehicles on the other hand, including hybrid, plug-in hybrid and battery electric vehicles, offer one fundamental advantage in this front. Through appropriate control strategy, the electric motor that drives the wheels can be used as a generator [3] that provides braking torque when the brakes are applied and also regenerates energy by converting the kinetic energy of the vehicle back into electrical energy which is then stored in the batteries. This technique can save considerable amount of energy increasing the fuel economy by reducing energy consumption and helps reduce emission of greenhouse gases [4].

Therefore to satisfy the features of regenerative braking as well as safety requirements, the brake system must provide a braking force that is a controllable combination of motor torque as well as mechanical frictional force. There are many architectures for achieving this [5], while the simplest one being parallel hybrid braking architecture [6].

Braking control strategy has two parts to it: (1) distribution of brake force between electric motor torque and mechanical friction forces, (2) distribution of forces to the front wheels and to the rear wheels (Section 2.1). In existing literature, this problem has been approached by using fuzzy logic in [7], [8] and [9] or designing specific solutions for a particular drive cycle like in [6]. Compared to the simple ANN model proposed in this paper that can solve the aforementioned problem, fuzzy logic is much more dependent on the membership functions one chooses to model. Designing specific solutions is not effective due to the variations in driving patterns. In this paper, we propose solving this distribution problem using an artificial neural network model in order to maximize the available energy for recovery while following the rules for stability. The ANN model is used on standard drive cycles [12] to get a quantitative estimate of available energy for recovery and comparison is provided. In the next section, we introduce the proposed methodology and then follow with the results and finally discuss conclusions.

2   Methodology

2.1   Problem Definition

Braking theory and design principle of conventional vehicle using frictional mechanism have been well established [6], which emphasizes distribution of total braking force on the front and rear wheels in order to obtain short braking distance and prevent the rear wheel being locked earlier than the front wheel locked in order to maintain the vehicle directional stability. Further, in an electric powertrain, we not only have to distribute brake forces between the front and rear axles but also between mechanical friction braking and electrical regenerative braking.

In this paper, we consider a hypothetical vehicle powertrain with the following parameters [6] in order make quantitative estimates about braking power. There are many architectures for the braking system [5]. In this paper we employ a parallel hybrid braking system because it has a simple structure and control. It retains the mechanical brake and adds electric regenerative brake on the front wheels. The electric motor that drives the front wheels, also applies brake torque directly through the front axle. Mechanical brake system consists of brake pedal and friction brakes and it requires a decoupling of the mechanical hydraulic connection between the brake pedal and the friction brake so that both the mechanical and the electrical brakes can be parally applied [10].

Table 1. Vehicle parameters used in this paper [6]

Parameters	Symbol	Unit	Value
Vehicle mass	M	Kg	1500
Rolling resistance coefficient	Fr		0.01
Aerodynamic drag coefficient	Cd		0.3
Front area	A	m^2	2.2
Wheel base	L	m	2.7
Gravity center height	h_g	m	0.6
Distance of gravity center to rear wheel center	L_b	m	1.89
Distance of gravity center to front wheel center	L_a	m	0.81

This way, the total braking force on the front wheels are just an addition of the individual braking forces.

$$F_{bf} = F_{fmech} + F_{regen} \tag{1}$$

where F_{bf} is the total braking force on front wheels,
    F_{fmech} is the friction braking force on front wheels, and
    F_{regen} is the regenerative electrical braking force on the front wheels.

Also, from the vehicle dynamics we have a few results:

$$F_{total} = F_{bf} + F_{br} \tag{2}$$

where F_{total} is the total braking force acting on the vehicle,
    F_{bf} is the total braking force on the front wheels and
    F_{br} is the total braking force on the rear wheels.
Also, for stability we have the below ratio [6],

$$\frac{F_{bf}}{F_{br}} = \frac{L_b + \frac{j}{g} * h_g}{L_a - \frac{j}{g} * h_g} \tag{3}$$

where j is the deceleration value,
    g is the acceleration due to gravity.
From these conditions, we have an ideal distribution of braking force between the front and rear axle so that both the axles lock simultaneously is plotted by varying the deceleration parameter from 0 to 1.2 g.

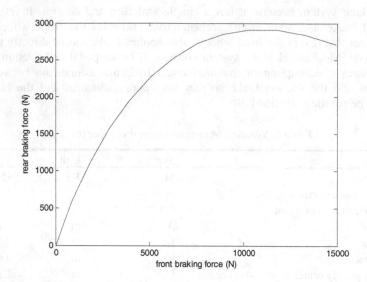

Fig. 1. Ideal brake distribution curve obtained from Eq (3)

Regenerative braking force F_{regen} can be given as follows:

$$F_{regen} = K * F_{bf} \qquad (4)$$

where K is defined as the portion of the front wheel braking force that is regenerative braking. Here, we consider K to be a function of deceleration j and velocity of the vehicle. And K can take the value between 0 and 1.

Further, F_{bf} is a portion of F_{total} and this portion is a function of deceleration (define it as Φ)

$$F_{bf} = \Phi * F_{total} \qquad (5)$$

Where (From (2) and (3)), $\Phi = \dfrac{1}{1+\dfrac{L_b+\frac{j}{g}*h_g}{L_a-\frac{j}{g}*h_g}}$

Therefore we have,

$$F_{regen} = K * \Phi * F_{total} \qquad (6)$$

Multiplying equation (6) both sides by velocity v, we get,

$$P_{regen} = K * \Phi * M * j * v \qquad (7)$$

$$P_{regen} = \dfrac{K * M * j * v}{1 + \dfrac{L_b + \dfrac{j}{g} * h_g}{L_a - \dfrac{j}{g} * h_g}} \qquad (8)$$

2.2   Applying Artificial Neural Networks

Artificial neural networks can be used to model non-linear equations successfully [11]. As it can be seen from Eq. (8), the brake force distribution is a non-linear function dependent on many parameters such as deceleration, velocity, and K function. We have considered modeling the K function using a neural network. In this paper we have used the MATLAB neural network toolbox for this purpose using deceleration and velocity as input parameters and the value of K as the output. Constraints are applied on maximum P_{regen} to generate training data for the neural network as described in section 2.3. Table 2 describes the architecture of the ANN model and the properties of the algorithm used.

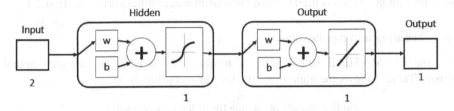

Fig. 2. The neural network model used in MATLAB

Fig. 2 shows the pictorial representation of the neural network architecture. W stands for weights vector and b stands for bias of the neuron. It has a 2x1 input vector with one neuron in the hidden layer with the sigmoid activation function and one in the output layer with a linear activation function. It gives the K function value as the only output. Single hidden neuron is found to be enough to represent the complexity of the system after some trial and error.

Table 2. Parameters for neural network model

Parameter	Value
Hidden Layer Size	1
Input Layer Size	2
Output Layer Size	1
Division of Data for Training	70/100
Division of Data for Validation	15/100
Division of Data for Testing	15/100
Training algorithm	Levenberg-Marquardt
Minimum Gradient	1.00e-05

2.3   Generating Training Data

From the Eq. (7), we have a definition for the K function. In order to train the neural network with input-output data, we use the following rules to generate data which maximizes the energy available for recovery. Here let us assume that the generator/motor is rated at 10 kW and has a maximum torque capability to support maximum deceleration, therefore the rules will be as follows:

1. Calculate K, for different data points (i.e, deceleration , velocity data pairs)
$$K = 10000/(M * j * v * \Phi)$$

2. If value of K is beyond the limit (i.e, $K > 1$), then it should be reset to 1, this means that we are including all available power below 10 kW

3. If value of K turns out to be less than 1, then K is set to that value for that data pair.

4. From this we get a K function that can be constructed using a neural network model with inputs as deceleration and velocity.

Next, the data generated is used to train the artificial neural network in section 2.4

2.4   Training and Application

We used the MATLAB neural network toolbox[13] to train the artificial neural network Table 3 shows the training results for various drive cycles.

Table 3. Results of training for different drive cycles

Drive cycles	Epoch (iterations)	Gradient	Performance (Mean square error)	Best validation performance
US_06	7	6.70e-06	0.00884	0.006652
US_06_HWY	7	3.12e-06	0.00410	0.002383
UDDS	9	5.87e-06	0.00064	0.001096

It can be seen that even a simple neural network model with a single neuron in the hidden layer can give us remarkable results in terms of final gradient after training. A gradient close to zero implies that the fitting error is very close to the minimum possible.

3   Results and Discussion

3.1   K Function Generation

Using neural network approach, a K function is generated and plotted using the deceleration and velocity data points from multiple drive cycle data such as UDDS, HWFET and US06. In the Figures 3, 4 and 5 we have velocity (in ms^{-1}) on the X axis, deceleration (in ms^{-2}) on the Y axis and the K function on the Z axis.

Fig 3 shows the K function map for the US06 drive cycle. It can be seen that the value of K function is 1 when the acceleration and velocity are closer to 0. This means that the entire braking torque on the front wheel is primarily regenerative braking torque. As we move to regions of higher velocity and deceleration, the value of K decreases in a hyperbolic fashion, this shows that the ratio of regenerative braking torque to the entire torque being applied to the front wheels has decreased and a considerable portion is friction torque. This shows that the motor is operating as a generator at its rated power and excess power is dissipated as heat. Similarly, Fig. 4 shows the K function map for the UDDS drive cycle and Fig. 5 shows the K function map for the US06HWY drive cycle.

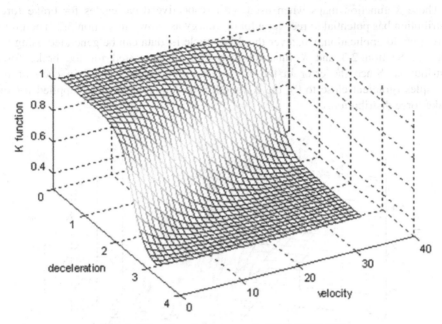

Fig. 3. *K* function for drive cycle CYC_US06 [12]

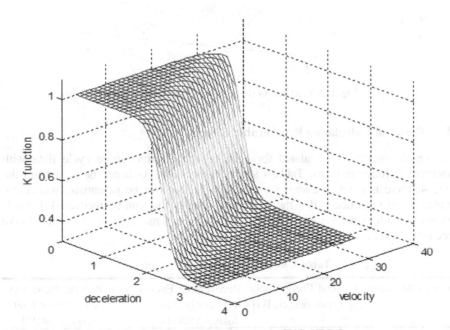

Fig. 4. *K* function for drive cycle CYC_UDDC

These K function maps when used with respective drive cycles for brake force distribution has potential to recover a lot of energy as shown in section 3.2. Therefore, in real world application, whatever the drive cycle is, data can be generated using the rules in Section 2.3 and K function map can be generated for the brake force distribution. Since the error in the ANN algorithm is very small in each of the examples (gradient close to 0), the K function is very reliable to be applied for the brake force distribution.

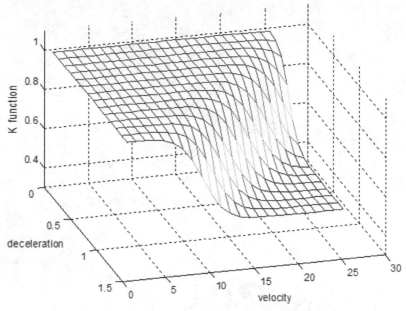

Fig. 5. K function for drive cycle CYC_US06 HWY

3.2   Result of Calculating Recoverable Energy

Recoverable energy is calculated for each of the different drive cycle data with respective K function maps. Table 4 shows the results. As it can be seen from the Table 4, there is a lot of energy (around 30 %) that can be potentially recovered. Another point to notice is that the amount of energy that can be recovered depends very much on how the vehicle is driven although the same logic is applied for brake force distribution.

Table 4. Results for various drive cycles

Drive cycle data used	Total Energy spent per drive cycle [kJ]	Recoverable Energy with regenerative braking (using ANN) [kJ]	Recoverable energy percentage of total energy spent [%]
CYC_US06	8.117	2.378	29.3
CYC_US06_HWY	5.358	1.817	33.9
CYC_ UDDS	5.129	1.882	36.7

This gives us more confidence that when the brake force distribution is applied to vehicles driven in urban areas with intermittent start-stop motion, for example the UDDS cycle, the percentage of energy recovery is considerably high. So any similar urban drive cycles will give us similar results with recoverable energy in the order of 35 % which is a huge saving in energy. This would result in lesser energy needed for commutation and decrease the size of battery required as expected.

4 Conclusions and Future Work

In this paper, a neural network based approach is introduced to determine the optimal brake force distribution between friction-regenerative and rear-front axle of an electrified vehicle powertrain. The percentage of energy recoverable varies for different drive cycles. For urban drive cycle such as UDDS, it was found that with the proposed approach around 37 % of the energy could be recovered per cycle while for an aggressive driving cycle such as US-06 the rate remains at 29%. An effective control system to recover energy will help improve fuel economy, reduce emissions and increase extended range for electric vehicles.

Though in this paper we have considered the parameters of deceleration and velocity to construct a K function model, there are many additional parameters to consider in practical applications such as the temperature of motor and state of charge for the battery system. Using a similar approach that is proposed in this paper, a more complex neural network model including these input parameters has the potential to perform well for the aforementioned objectives as well as safety of vehicle operation.

References

1. Shakouri, P., Ordys, A., Askari, M., Laila, D.S.: Longitudinal vehicle dynamics using Simulink/Matlab. In: UKACC International Conference on Control 2010, September 7-10, pp. 1–6 (2010), doi:10.1049/ic.2010.0410
2. Lintern, M.A., Chen, R., Carroll, S., Walsh, C.: Simulation study on the measured difference in fuel consumption between real-world driving and ECE-15 of a hybrid electric vehicle. In: Hybrid and Electric Vehicles Conference 2013 (HEVC 2013), November 6-7, pp. 1–6. IET (2013)
3. Cao, B., Bai, Z., Zhang, W.: Research on control for regenerative braking of electric vehicle. In: IEEE International Conference on Vehicular Electronics and Safety, October 14-16, pp. 92–97 (2005), doi:10.1109/ICVES.2005.1563620
4. Ortmeyer, T.H., Pillay, P.: Transportation sector technology energy use and GHG emissions. In: 2002 IEEE Power Engineering Society Summer Meeting, July 25-25, vol. 1, pp. 34–35 (2002), doi:10.1109/PESS.2002.1043173
5. Mutoh, N., Hayano, Y., Yahagi, H., Takita, K.: Electric Braking Control Methods for Electric Vehicles With Independently Driven Front and Rear Wheels. IEEE Transactions on Industrial Electronics 54(2), 1168–1176 (2007), doi:10.1109/TIE.2007.892731
6. Gao, Y., Chu, L., Ehsani, M.: Design and Control Principles of Hybrid Braking System for EV, HEV and FCV. In: IEEE Vehicle Power and Propulsion Conference, VPPC 2007, September 9-12, pp. 384–391 (2007)

7. Xu, G., Li, W., Xu, K., Song, Z.: An Intelligent Regenerative Braking Strategy for Electric Vehicles. . Energies 2011 4, 1461–1477 (2011)
8. Jing-Ming, Z., Bao-Yu, S., Shu-Mei, C., Dian-Bo, R.: Fuzzy Logic Approach to Regenerative Braking System. In: International Conference on Intelligent Human-Machine Systems and Cybernetics, IHMSC 2009, August 26-27, vol. 1, pp. 451–454 (2009)
9. Bathaee, S.M.T., Gastaj, A.H., Emami, S.R., Mohammadian, M.: A fuzzy-based supervisory robust control for parallel hybrid electric vehicles. In: 2005 IEEE Conference on Vehicle Power and Propulsion, September 7-9, p. 7 (2005)
10. Zeng, X., Ba, T., Wang, Q., Qu, X., Song, D.: A kind of accurately Optimized braking energy distribution strategy applied to switched Series-parallel Hybrid Electric Bus. In: 2011 2nd International Conference on Artificial Intelligence, Management Science and Electronic Commerce (AIMSEC), August 8-10, pp. 3634–3637 (2011)
11. Li, G., Zeng, Z.: A Neural-Network Algorithm for Solving Nonlinear Equation Systems. In: International Conference on Computational Intelligence and Security, CIS 2008, December 13-17, vol. 1, pp. 20–23 (2008), doi:10.1109/CIS.2008.65
12. National Renewable Energy Laboratory. ADVISOR Documentation, [EB/OL] (2001-01-19) [2005-04-15], http://www.ctts.nrel.gov/analysis/
13. Neural Network Toolbox [MathWorks MATLAB], http://www.mathworks.com/products/neural-network/

Automatic Screening and Classification of Diabetic Retinopathy Fundus Images

Sarni Suhaila Rahim, Vasile Palade, James Shuttleworth, and Chrisina Jayne

Faculty of Engineering and Computing, Coventry University, Priory Street, Coventry,
CV1 5FB United Kingdom
rahims3@uni.coventry.ac.uk
{vasile.palade,csx239,ab1527}@coventry.ac.uk

Abstract. Eye screening is essential for the early detection and treatment of the diabetic retinopathy. This paper presents an automatic screening system for diabetic retinopathy to be used in the field of retinal ophthalmology. The paper first explores the existing systems and applications related to diabetic retinopathy screening and detection methods that have been previously reported in the literature. The proposed ophthalmic decision support system consists of an automatic acquisition, screening and classification of diabetic retinopathy fundus images, which will assist in the detection and management of the diabetic retinopathy. The developed system contains four main parts, namely the image acquisition, the image preprocessing, the feature extraction, and the classification by using several machine learning techniques.

Keywords: Diabetic Retinopathy, Eye Screening, Eye Fundus Images, Image Processing, Classifiers.

1   Introduction

Screening is defined as testing on a population in order to identify individuals exhibiting attributes that could be early symptoms or indicators of predisposition associated with a particular condition. Screening is used to maximise the chances of any individual overcoming the threat or danger indicated by such attributes (Taylor and Batey, 2012).

The main purpose of diabetic retinopathy screening is to detect whether the individuals require follow up or referral for further treatment to prevent blindness (Taylor and Batey, 2012). Besides this main purpose, there are other purposes for diabetic retinopathy screening, which include: identifying the disease at an early stage; possibly detecting a requirement for blood pressure and blood sugar treatment; to educate the population on the diabetic retinopathy causes and on the ways to reduce the retinopathy risk; and, additionally, to potentially identify non-diabetic conditions through the screening process.

Diabetes Mellitus (DM) is a major public health concern, as it leads to an increasing number of acute and chronic complications, including sight-threatening conditions. Diabetic Retinopathy (DR) is one of the chronic complications of

V. Mladenov et al. (Eds.): EANN 2014, CCIS 459, pp. 113–122, 2014.

diabetes, and it is a microvascular complication of both insulin dependent (type 1) and non-insulin dependent (type 2) diabetes. DR is a complication of DM that damages blood vessels inside the retina at the back of the eye. Wild and co-workers (2004) revealed that the global prevalence of diabetes mellitus in 2000 was approximately 2.8% (171 million diabetics) and projected this to rise to 4.4% (366 million diabetics) in 2030. According to Taylor and Batey (2012), one major problem is that the diabetic eye disease does not interfere with sight until it reaches an advanced stage. Laser treatment can save sight, but only if it is used at an early stage and, hence, regular screening is essential. This shows the importance of regular screening, which can help detect the diabetic patients at an early stage of DR. Furthermore, earlier identification of any retinopathy can allow change in blood pressure or blood glucose management to slow the rate of the disease progression.

2   Existing Systems

There currently are several developed systems to detect and diagnose diabetic retinopathy (DR), and most of these existing systems are somewhat related to the proposed system and can be used as a benchmark. Diabetic retinopathy screening is a popular research area and a lot of researchers focus on and contribute towards the advancement of study in this area.

Some of them focused on finding and proposing an accurate technique or method for detecting certain features of DR fundus images, such as microaneurysms, hemorrhages and neovascularisation. An automated grading system with image processing methods that detect two DR features, which are the dot hemorrhages and microaneurysms, was developed by Larsen and colleagues (2002). Jelinek and others (2006) developed an effective tool for detecting microaneurysms, in order to identify the DR presence in rural optometric practices. A comparison of the automated system used with optometric and ophthalmologic assessment was performed by calculating the sensitivity and specificity of both methods.

Nonetheless, there are some researches that report the development of automated systems for detecting DR by classifying DR into general detection categories, such as normal (no apparent retinopathy) or abnormal (retinopathy presence). Also, there are other classification systems that provide more details about the retinopathy stages, which include normal, non-proliferative diabetic retinopathy (NPDR) and proliferative diabetic retinopathy (PDR). Priya and Aruna (2011) investigated and proposed a computer-based system for identifying normal, NPDR and PDR classes. The proposed system uses colour fundus images, where the features are extracted from the raw image with image processing techniques and fed to a Support Vector Machine (SVM) for classification. The system has been later enhanced by using two types of classifiers, a Probabilistic Neural Network (PNN) and a Support Vector Machine (Priya and Aruna, 2012). The classifiers are described in detail and their performances are compared. As a conclusion, it is shown that, from the results obtained, the SVM model is more effective compared to the PNN. Priya and Aruna (2013) proposed and compared three models, i.e., a Bayesian classifier while maintaining the PNN and the SVM in the developed system. Experimental results show that the SVM outperforms all other models and this proves, once again, that the SVM is a great choice to use in detecting and classifying DR categories. The detection of the DR disease and the classification with the help of Radial Basis Function Neural Network (RBFNN) method has been proposed in (Priya *et al*, 2013). However, the experimental results show that the accuracy of the proposed system is relatively low, (76.25%), and it is recommended that this method could be improved by finding more relevant features and by combining with other classification methods, in order to improve the accuracy rate. The Aravind Diabetic Retinopathy Screening (ADRES) 3.0, developed and presented by Permalsamy and colleagues (2007), is a software for reading and grading the DR. This simple tool is used to assist in the detection of the DR and it is offered as a supplementary checking method to an usual clinical examination by an ophthalmologist. Philip et al. (2007) presented a study on the efficiency of the manual versus automated "disease" or "no disease" grading systems against the reference standard.

3 Proposed System

In this paper, an automatic classification and screening of diabetic retinopathy (DR) using fundus images is presented. A combination of normal and DR affected fundus images from a public database, i.e., the Standard Diabetic Retinopathy Database Calibration Level 0 (DIARETDB0), have been used for the evaluation of the proposed system. The database consists of 130 colour fundus images, of which 20 are normal and 110 contain signs of diabetic retinopathy (hard exudates, soft exudates, microaneurysms, hemorrhages and neovascularisation). The original images, which are of size 1500 x 1152 in PNG format, were captured with a 50 degree field-of-view digital fundus cameras with unknown camera settings (Kauppi *et al*, 2006). The proposed screening system has been developed using open source software, OpenCV (Open Source Computer Vision) and Microsoft Visual C++ 2010. The OpenCV environment, developed by Willow Garage, is a programming library offered for real time computer vision (Itseez, 2014). OpenCV includes a collection of standardised image analysis and machine vision algorithms for use by developers. Most work in the area has used tools such as Matlab and SPSS for feature extraction and analysis, but by using OpenCV it is possible to build more efficient systems, with processing times suitable for use in real situations. Using OpenCV also simplifies the distribution of software due to permissive licensing, and it lowers the cost of development, use and maintenance because there are no purchase or licensing fees. Finally, OpenCV is portable, meaning that any machine that can run C can, most likely, also run OpenCV. OpenCV has been used on Windows, Linux, MacOS and Android, for example.

The proposed system starts with the image acquisition process, where the system will select images for further processing. The selected images will undergo preprocessing in order to improve the image contrast as well as perform other enhancements. After that, the preprocessed images will be used to extract a number of features, such as the area, the mean and the standard deviation of on pixels.

Four nonlinear classifiers, namely a binary decision tree, a k-nearest neighbour classifier, and two support vector machines, using radial basis function and polynomial function kernels, respectively, are then trained on the training data to find an optimal way to classify images into their respective classes. Finally, in the prediction phase, where the system might ultimately be used to help the clinician, the images are classified into two main groups: normal or DR.

The remainder of this paper is organised as follows. Section 4 describes the image preprocessing stage followed by Section 5, which explains the feature extraction part. Section 6 describes the nonlinear classification, while Section 7 presents the results of the system and, finally, Section 8 details the conclusions of the work and a future plan.

Figure 1 presents the block diagram of the proposed system for automating the screening and classification of the diabetic retinopathy. Individual stages will be discussed in more detail in the following sections.

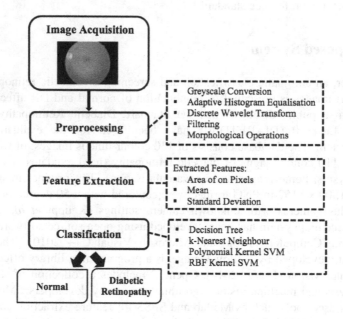

Fig. 1. Block diagram of the proposed automatic screening and classification of diabetic retinopathy

4   Image Preprocessing

Preprocessing is the process of image data improvement, where enhancing some image characteristics/features for the next processing part takes place. The image preprocessing techniques involved in the present work include Greyscale Conversion, Adaptive Histogram Equalisation, Discrete Wavelet Transform, Filtering and Morphological Operations.

4.1   Greyscale Conversion

The first preprocessing technique used is converting the colour fundus image into a greyscale image, as greyscale is usually the ideal format for image processing. A greyscale image is an image where each pixel holds a single value, only the pixel intensity information. It is also known as "black and white" image. The intensity is calculated by using a common formula combination of 30% of red, 59% of green and 11% of blue.

4.2   Adaptive Histogram Equalisation

Adaptive Histogram Equalisation (AHE) is a computer image processing technique for improving the image's contrast. The difference between the adaptive histogram equalisation and the ordinary histogram equalisation is that the adaptive histogram equalisation computes several histograms, for different sections of the image, and subsequently distributes the lightness values. This technique is used to improve the local contrast and bringing out more details of the image. However, the adaptive histogram equalisation has limitations, as it produces over-amplification of noise in the homogeneous regions of an image. Therefore, the Contrast Limited Adaptive Histogram Equalisation (CLAHE) is used in the proposed system in order to prevent the overamplification of noise. CLAHE functions by clipping the histogram at the predefined value before computing the cumulative distribution function.

4.3   Discrete Wavelet Transform

Discrete wavelet transform is the discrete variant of the wavelet transform. The discrete wavelet transform is an $O(N)$ algorithm and it is also often referred to as the fast wavelet transform. The Haar wavelet is implemented in the proposed system development as it is a simple wavelet transform and it is being used in many methods of discrete image transforms and processing. Discrete wavelet transforms can be used to reduce the image size without losing much of the resolution. Since the fundus images are of high resolution and of quite large size, the Haar wavelet is recommended to be used.

4.4   Filtering

Image filtering is used to improve the image quality or restore the digital image which has been corrupted by some noise. A comparison of the performance between three different edge operators, i.e., Sobel, Prewitt and Kirsch has been proposed for the detection and segmentation of blood vessels in the colour retinal images (Karasulu, 2012). The experimental results show that the edge-based segmentation using Kirsch compass templates is superior by far to other methods. Based on this result, the Kirsch operator has been chosen for filtering in the proposed system development. The Kirsch edge detection uses eight filters, which means eight masks for the related eight main directions are applied to a given image to detect edges. These eight filters are a rotation of a basic 3x3 compass convolution filter. The Kirsch filter is applied on the wavelet transform image to create the eight filtered output image.

4.5   Morphological Operations

Morphological operations are used for certain purposes including the image preprocessing, enhancing object structure, segmenting objects from the background and also for quantitative description of objects (Sonka *et al*, 2008). In the proposed system development, morphology operators involving dilation and erosion are implemented to extract the blood vessels. A closing operation is defined as dilation followed by erosion operator. Joshi and Karule (2012) implemented the closing operation for retinal blood vessel segmentation, where the disk shaped structuring element for morphological operation is used. The dilation operates in greyscale images to enlarge brighter regions and it closes the small dark regions, while the erosion operator shrinks the dilated objects back to the original size and shape. As a result, the vessels being thin dark segments laid out on a brighter background are closed by the closing operation. Figure 2 (a)-(f) shows the output after each of the preprocessing operations on an image selected, as explained previously.

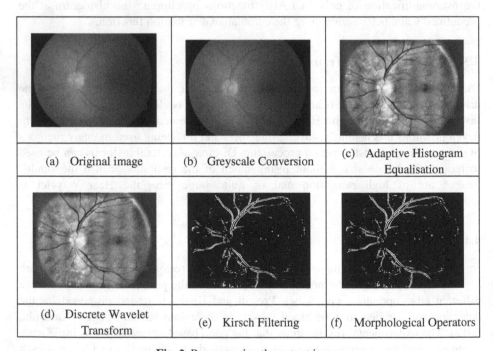

Fig. 2. Preprocessing the output image

5   Feature Extraction

After performing the preprocessing techniques, feature extraction takes place in order to obtain the features from the given images. Features such as the area of on pixels, mean and standard deviation are extracted for diabetic retinopathy (DR) detection purposes. These values for both normal and DR images are used to create a model for training. Table 1 presents the details of the feature extracted including the generated code.

Table 1. Feature extraction in the proposed system

Feature	Description	Snippet Code

Area of on pixels	Number of white pixels on the black and white image, where white pixels on are all pixels above a threshold of 100.	```maxS = cvGet2D(gray, y, x);```
		```val = maxS.val[0];```
		```if(val > 100) {```
		``` count++;```
		``` sum += val; }```

Mean	Mean value of on pixels	```mean = sum / count;```

Standard deviation	Standard deviation of on pixels	```maxS = cvGet2D(gray, y, x);```
		```val = maxS.val[0];```
		```if(val > 100) {```
		``` count++;```
		``` val = (mean-val);```
		``` sum += val*val; }```
		```sdv = sqrt(sum) / count;```

6   Classification

The feature extracted values from the developed system have been passed to Matlab for the classification stage in order to benefit from various classifiers available in Matlab. The PRTools, a Matlab toolbox for pattern recognition has been downloaded and used in Matlab (Duin *et al*, 2007). Nonlinear classifiers can provide better classification results compared to linear classifiers. Therefore, four nonlinear classifiers, namely the binary decision tree classifier, the k-nearest neighbour classifier, the RBF kernel based support vector classifier and the polynomial kernel based support vector classifier have been selected to train and classify images into two classes, i.e., normal and diabetic retinopathy, respectively, based on the three extracted features as explained previously in Section 5. Decision tree is a classifier in the form of a tree structure and classifies instances or examples by starting at the root of the tree and moving through it until a leaf node is reached. In the k-nearest neighbour classifier, the object is classified by a majority vote of its neighbours, with the object being assigned to the most common class among its *k* nearest neighbours. The 1-nearest neighbour rule (1-NN) is used in the particular implementation of the system presented in the paper. A support vector machine (SVM) performs the classification by constructing an *N*-dimensional hyperplane that optimally separates the data into two categories. The support vector machine classifier can use various kernel functions, such as linear, polynomial or radial basis function (RBF). The kernel function transforms the data into a higher dimensional space in order to be able to perform the separation in the nonlinear region. Two different types of kernel functions provided in Matlab for SVM classification were used, i.e., the second order polynomial kernel SVM, *svc (ATrain, 'p', 2)*, and the radial basis function kernel SVM, *rbsvc (ATrain)*. The results show that RBF kernel outperformed the results obtained with the second order polynomial kernel.

7   Results and Discussion

Figure 3 shows the user interface snapshot of the proposed developed system. The performance (misclassification error) of the four classifiers is presented in Table 2. Since the dataset is hugely unbalanced, the minority class was oversampled by duplication in order to balance the dataset. The DIARETDB0 data is split randomly into 90% for training and the remaining 10% for testing. The process is repeated ten times in a cross-validation procedure in order to generate unbiased results. The average results on the ten runs for each of the four classifiers are reported. For more clarity, in Table 2, we also presented the confusion matrix for the first out of the ten experiments, in order to show the relative performance of the four classifiers. The classification performance of the diagnosis system is assessed using the accuracy of the individual classifiers and also the specificity and sensitivity. The experimental results show that the four classifiers, and especially the k- nearest neighbor, are able to identify well both classes, i.e., the normal and the diabetic retinopathy cases. All the four classifiers identified much better the diabetic retinopathy cases, as there were more examples of such images in the database.

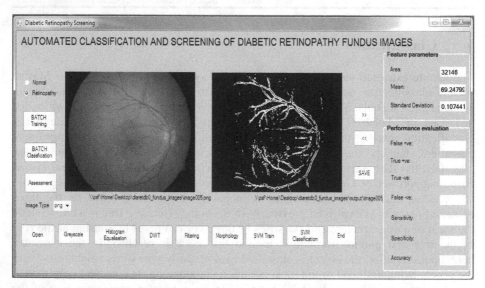

Fig. 3. Snapshot of the proposed system user interface

Table 2. Average results when using the four classifiers

	Binary decision tree	k-nearest neighbour	RBF kernel SVM	Polynomial kernel SVM
Misclassification error	0.2091	0.01364	0.0909	0.3182
Accuracy	0.7909	0.9864	0.9091	0.6818
Specificity	1	1	1	0.5545
Sensitivity	0.5818	0.9727	0.8182	0.8091

Confusion matrix for the first experiment Labels (1 : Normal, 2: DR)

Binary decision tree:
```
True   | Estimated Labels
Labels | 1   2 | Totals
-------|-------|--------
1      | 11  0 | 11
2      | 5   6 | 11
-------|-------|--------
Totals | 16  6 | 22
```

k-nearest neighbour:
```
True   | Estimated Labels
Labels | 1   2 | Totals
-------|-------|--------
1      | 11  0 | 11
2      | 0  11 | 11
-------|-------|--------
Totals | 11 11 | 22
```

RBF kernel SVM:
```
True   | Estimated Labels
Labels | 1   2 | Totals
-------|-------|--------
1      | 11  0 | 11
2      | 2   9 | 11
-------|-------|--------
Totals | 13  9 | 22
```

Polynomial kernel SVM:
```
True   | Estimated Labels
Labels | 1   2 | Totals
-------|-------|--------
1      | 6   5 | 11
2      | 2   9 | 11
-------|-------|--------
Totals | 8  14 | 22
```

8 Conclusions and Future Work

An automatic system for screening and classification of the diabetic retinopathy (DR) using fundus images has been developed. The system will be enhanced on the classification part by building an ensemble of classifiers. Unbalanced learning techniques will also be considered to be used for training the individual classifiers in the ensemble. In addition, more sophisticated features will be used in our future work to properly discriminate the various diabetic retinopathy signs (i.e., different features extraction for microaneurysms, hemorrhages, exudates, etc.). The system will also be extended to get more details on the DR classification, namely to classify into no apparent retinopathy, mild non-proliferative, moderate non-proliferative, severe non-proliferative and proliferative DR cases. In addition to the classification diagnosis, the system will provide the recommended follow-up schedule for each stage, as underlined by the American Academy of Ophthalmology and, hence, this will become a complete system to be used in a diabetic retinopathy screening practice.

Acknowledgement. This project is a part of PhD research currently being carried out at the Faculty of Engineering and Computing, Coventry University, United Kingdom. The deepest gratitude and thanks go to the Universiti Teknikal Malaysia Melaka (UTeM) and Ministry of Education Malaysia for sponsoring this PhD research.

References

1. Duin, R.P.W., Juszczak, P., Paclik, P., Pekalska, E., de Ridder, D., Tax, D.M.J., Verzakov, S.: PRTools4.1, A Matlab Toolbox for Pattern Recognition, Delft University of Technology (2007)
2. Itseez, http://opencv.org
3. Jelinek, H.J., Cree, M.J., Worsley, D., Luckie, A., Nixon, P.: An automated microaneursym detector as a tool for identification of diabetic retinopathy in rural optometric practice. Clinical and Experimental Optometry 89(5), 299–305 (2006)
4. Joshi, S., Karule, P.T.: Retinal blood vessel segmentation. International Journal of Engineering and Innovative Technology 1(3), 175–178 (2012)

5. Karasulu, B.: Automated extraction of retinal blood vessels: a software implementation. European Scientific Journal 8(30), 47–57 (2012)
6. Kauppi, T., Kalesnykiene, V., Kamarainen, J.-K., Lensu, L., Sorri, I., Uusitalo, H., Kalviainen, H., Pietila, J.: DIARETDB0: Evaluation Database and Methodology for Diabetic Retinopathy Algorithms, Technical report (2006)
7. Larsen, N., Godt, J., Grunkin, M., Lund-Andersen, H., Larsen, M.: Automated detection of diabetic retinopathy in a fundus photographic screening population. Investigative Ophthalmology and Visual Science 44(2), 767–771 (2003)
8. Perumalsamy, N., Sathya, S., Prasad, N.M., Ramasamy, K.: Software for reading and grading diabetic retinopathy. Aravind Diabetic Retinopathy Screening 3.0 Diabetes Care 30(9), 2302–2306 (2007)
9. Philip, S., Fleming, A.D., Goatman, K.A., Fonseca, S., Mcnamee, P., Scotland, G.S., Sharp, P.F., Olson, J.A.: The efficacy of automated "disease/no disease" grading for diabetic retinopathy in a systematic screening programme. British Journal Ophthalmol. 91, 1512–1517 (2007)
10. Priya, R., Aruna, P.: Review of automated diagnosis of diabetic retinopathy using the support vector machine. International Journal of Applied Engineering Research 1(4), 844–863 (2011)
11. Priya, R., Aruna, P.: SVM and neural network based diagnosis of diabetic retinopathy. International Journal of Computer Applications 41(1), 6–12 (2012)
12. Priya, R., Aruna, P.: Diagnosis of diabetic retinopathy using machine learning techniques. Journal on Soft Computing 3(4), 563–575 (2013)
13. Priya, R., Aruna, P., Suriya, R.: Image analysis technique for detecting diabetic retinopathy. International Journal of Computer Applications 1, 34–38 (2013)
14. Sonka, M., Hlavac, V., Boyle, R.: Image Processing, Analysis, and Machine Vision. Cengage Learning. United States of America (2008)
15. Taylor, R., Batey, D.: Handbook of retinal screening in diabetes: diagnosis and management. John Wiley & Sons, Ltd., England (2012)
16. Wild, S., Roglic, G., Green, A., Sicree, R., King, H.: Global prevalence of diabetes estimates: estimates for the year 2000 and projections for 2030. Diabetes Care 27(5), 1047–1053 (2004)

Brain Neural Data Analysis Using Machine Learning Feature Selection and Classification Methods

Lachezar Bozhkov[1], Petia Georgieva[2], and Roumen Trifonov[1],

[1] Computer Systems Department, Technical University of Sofia, 8 St.Kliment Ohridski Boulevard,Sofia 1756, Bulgaria
[2] DETI/IEETA, University of Aveiro, 3810-193 Aveiro, Portugal
lachezar.bozhkov@gmail.com, petia@ua.pt, r_trifonov@tu-sofia.bg

Abstract. The Electroencephalogram (EEG) is a powerful instrument to collect vast quantities of data about human brain activity. A typical EEG experiment can produce a two-dimensional data matrix related to the human neuronal activity every millisecond, projected on the head surface at a spatial resolution of a few centimeters. As in other modern empirical sciences, the EEG instrumentation has led to a flood of data and a corresponding need for new data analysis methods. This paper summarizes the results of applying supervised machine learning (ML) methods to the problem of classifying emotional states of human subjects based on EEG. In particular, we compare six ML algorithms to distinguish event-related potentials, associated with the processing of different emotional valences, collected while subjects were viewing high arousal images with positive or negative emotional content. 98% inter-subject classification accuracy based on the majority of votes between all classifiers is the main achievement of this paper, which outperforms previous published results.

Keywords: emotion valence recognition, feature selection, Event Related Potentials (ERPs).

1   Introduction

The quantification and automatic detection of human emotions is the focus of the interdisciplinary research field of Affective Computing (AC). In [1] a broad overview of the current AC systems is provided. Major modalities for affect detection are facial expressions, voice, text, body language and posture. Affective neuroscience is a new modality that attempt to find the neural correlates of emotional processes [2]. Literature on learning to decode human emotions from Event Related Potentials (ERPs) was reviewed by [3], building automatic recognition systems from EEG was proposed by [4] and [5]. Despite the first promising results of the affective neuroscience modality to decode basic human emotional states, a confident neural model of emotions is still not defined. The recent overview of EEG-based emotion recognition studies, provided in [6], show that the recognition rate ranges between 65-90 %. Therefore, the primary motivation of the present paper is to determine a

V. Mladenov et al. (Eds.): EANN 2014, CCIS 459, pp. 123–132, 2014.

framework to improve the recognition of human affective states based on brain data and more particularly on ERPs. ERPs are transient components in the EEG generated in response to a stimulus (a visual or auditory stimulus, for example). We studied six supervised machine learning (ML) algorithms, namely Artificial Neural Networks (ANN), Logistic Regression (LogReg), Linear Discriminant Analysis (LDA), k-Nearest Neighbors (kNN), Naïve Bayes (NB), Support Vector Machines (SVM), Decision Trees (DT) and Decision Tree Bootstrap Aggregation (Tbagger) to distinguish affective valences encoded into the ERPs collected while subjects were viewing high arousal images with positive or negative emotional content. Our work is also inspired by advances in experimental psychology [7], [8] that show a clear relation between ERPs and visual stimuli with underlined negative content (images with fearful and disgusted faces). A crucial step preceding the classification process is to discover which spatial-temporal patterns (features) in the ERPs indicate that a subject is exposed to stimuli that induce emotions. We applied successfully the Sequential Feature Selection (SFS) technique to minimize significantly the number of the relevant spatial temporal patterns.

The paper is organized as follows. In section 2 we briefly describe the data set. The ML feature selection and classification methods used in this study are summarized in section 3. The results of learning to discriminate emotional states with positive or negative valences across multiple subjects (inter-subject setting) are presented in section 4. Finally, in section 5 our conclusions are drawn.

2   Data Set

A total of 26 female volunteers participated in the study, 21 channels of EEG, positioned according to the 10-20 system and 2 EOG channels (vertical and horizontal) were sampled at 1000Hz and stored. The signals were recorded while the volunteers were viewing pictures selected from the International Affective Picture System. A total of 24 of high arousal (> 6) images with positive valence (7.29 +/- 0.65) and negative valence (1.47 +/- 0.24) were selected. Each image was presented 3 times in a pseudo-random order and each trial lasted 3500ms: during the first 750ms, a fixation cross was presented, then one of the images during 500ms and at last a black screen during the 2250ms.

The signals were pre-processed (filtered, eye-movement corrected, baseline compensation and epoched using NeuroScan. The single-trial signal length is 950ms with 150ms before the stimulus onset. The ensemble average for each condition was also computed and filtered using a zero-phase filtering scheme. The maximum and minimum values of the ensemble average signals were detected. Then starting by the localization of the first minimum the features are defined as the latency and amplitude of the consecutive minimums and the consecutive maximums: minimums (Amin1, Amin2, Amin3), the first three maximums (Amax1, Amax2, Amax3), and their associated latencies (Lmin1, Lmin2, Lmin3, Lmax1, Lmax2, Lmax3). The ensemble average for each condition (positive/negative valence) was also computed and filtered using a Butterworth filter of 4th order with passband [0.5 - 15]Hz. The number of features stored per channel is 12 corresponding to the latency (time of occurrence) and amplitude of either n = 3 maximums and minimums, the features correspond to the time and amplitude characteristics of the first three minimums occurring after T = 0s and the corresponding maximums in between. The total number of features per trail is 252. The data is saved in file with the following structure: 252 columns: 12 features for 21 channels, 52 lines: 26 people x 2 classes – 0 (negative) and 1 (positive).

3   Classification Methodology

Predictor data is normalized to maximally ease the learning algorithms.. In order to maximize the training examples, leave-one-out cross-validation technique is used. The following supervised machine learning models are studied: Artificial Neural Networks (ANN), Logistic Regression (LogReg), Linear Discriminant Analysis (LDA), k-Nearest Neighbors (kNN), Naïve Bayes (NB), Support Vector Machines (SVM), Decision Trees (DT) and Decision Tree Bootstrap Aggregation (Tbagger).

3.1   Features Normalization

Many of the models require normalized version of the data. The rest of the models can highly benefit from it. Therefore this is often a good preprocessing practice.

Feature normalization is a standard preprocessing step, that may improve the classification, particularly when the range of the features is dispersed. There are a number of normalization techniques, in this work we use the following expression:

$$X_{norm} = (X - X_{mean}) / std(X) , \tag{1}$$

The normalized data (X_{norm}) is obtained by subtracting the mean value of each feature from the original data set X and divided by the standard deviation std(X). Hence, the normalized data has zero mean and standard deviation equal to 1.

3.2   Leave-One-Out Cross-Validation (LOOCV)

Leave-one-out is the degenerate case of K-Fold Cross Validation, where K is chosen as the total number of examples. For a dataset with N examples, perform N experiments. For each experiment use N-1 examples for training and the remaining 1 example for testing [9]. In our case N = 26 (pairs of classes per person). We will train the models with 25 people x 2 classes (50 examples) and test on the left-out 2 classes. We are more interested in the total prediction accuracy for each model, therefore the predictions are accumulated in confusion matrices for each model from each training experiment in the LOOCV.

3.3   Artificial Neural Network (ANN)

The ANNs origin from algorithms that try to mimic the brain neuronal structure. ANNs are widely used ML technique as classifiers and repressors in countless applications. In the present work, prediction is performed by a feedforward neural network (FFNN) with 1 hidden layer with 12 neurons with sigmoid activation function and training is performed by backpropagation algorithm to compute the gradient [10].

3.4   Logistic Regression (LogReg)

In statistics, LogReg is a type of probabilistic statistical classification model [11]. It is also used to predict a binary response from a binary predictor, used for predicting the outcome of a categorical dependent variable (i.e., a class label) based on one or more predictor variables (features).

3.5   Linear Discriminant Analysis (LDA)

Discriminant analysis is a classification method. It assumes that different classes generate data based on different Gaussian distributions. To train (create) a classifier, the fitting function estimates the parameters of a Gaussian distribution for each class. To predict the classes of new data, the trained classifier finds the class with the smallest misclassification cost. LDA is also known as the Fisher discriminant, named for its inventor, Sir R. A. Fisher [12].

3.6   k-nearest Neighbor (kNN)

Given a set X of n points and a distance function, kNN searches for the k closest points in X to a query point or set of points Y [13]. The kNN search technique and kNN-based algorithms are widely used as benchmark learning rules. The relative simplicity of the kNN search technique makes it easy to compare the results from other classification techniques to kNN results. The distance measure is Euclidean.

3.7   Naive Bayes (NB)

The NB classifier is designed for use when features are independent of one another within each class, but it appears to work well in practice even when that independence assumption is not valid. It classifies data in two steps:

Training step: Using the training samples, the method estimates the parameters of a probability distribution, assuming features are conditionally independent given the class.

Prediction step: For any unseen test sample, the method computes the posterior probability of that sample belonging to each class. The method then classifies the test sample according the largest posterior probability.

The class-conditional independence assumption greatly simplifies the training step since you can estimate the one-dimensional class-conditional density for each feature individually. While the class-conditional independence between features is not true in general, research shows that this optimistic assumption works well in practice. This assumption of class independence allows the NB classifier to better estimate the parameters required for accurate classification while using less training data than many other classifiers. This makes it particularly effective for datasets containing many predictors or features [13].

3.8   Support Vector Machines (SVM)

An SVM classifies data by finding the best hyperplane that separates all data points of one class from those of the other class. The best hyperplane for an SVM means the one with the largest margin between the two classes. Margin means the maximal width of the slab parallel to the hyperplane that has no interior data points. We use radial basis function for kernel function [13].

3.9   Decision Tree (DT)

Classification trees and regression trees are the two main DT techniques to predict responses to data. To predict a response, follow the decisions in the tree from the root (beginning) node down to a leaf node. The leaf node contains the response. Classification trees give responses that are nominal, such as 'true' or 'false' [13].

3.10   Decision Tree Bootstrap Aggregation (Tbagger)

Bagging, which stands for "bootstrap aggregation," is a type of ensemble learning. To bag a weak learner such as a decision tree on a dataset, generate many bootstrap replicas of this dataset and grow decision trees on these replicas. Obtain each bootstrap replica by randomly selecting N observations out of N with replacement, where N is the dataset size. To find the predicted response of a trained ensemble, take an average over predictions from individual trees [13].

4   Features Selection

The feature space consists of 252 features (21 channels x12 features) and the trial examples are 52 (2 classes x 26 people), therefore feature reduction techniques are required. First classification tests are made on all predictor data features (252 features) and the accuracy results from ML methods are set as base line to improve and compare. Next we try feature reduction using Principal Component Analysis (PCA) [14] and dimensions reduction with 99%, 95%, 75% and 50% data variation retained. After that we implement exhaustive feature selection and compare the results. Finally we construct voting ensemble bucket of models to take the prediction among all the models which resulted in very promising final data discrimination (98%).

4.1   Principal Component Analysis (SFS)

Principal component analysis is a quantitatively rigorous method for achieving this simplification. The method generates a new set of variables, called principal components. Each principal component is a linear combination of the original variables. All the principal components are orthogonal to each other, so there is no redundant information. The principal components as a whole form an orthogonal basis for the space of the data [13].

4.2   Sequential Feature Selection (SFS)

Sequential feature selection selects a subset of features from the data matrix X that best predict the data in y by sequentially selecting features until there is no improvement in prediction. Starting from an empty feature set, SFS creates candidate feature subsets by sequentially adding each of the features not yet selected. For each candidate feature subset, SFS performs leave-one-out cross-validation by repeatedly calling fun with different training subsets X_{TRAIN} and y_{train}, and test subsets X_{TEST} and y_{test}. Each time it is called, fun must return a scalar value criterion. After computing the mean criterion values for each candidate feature subset, SFS chooses the candidate feature subset that minimizes the mean criterion value. This process continues until adding more features does not decrease the criterion or to predefined number of selected feature. In our case the criterion function is based on the accuracy of the model: criterion = 1 – Accuracy. Accuracy can be either 1 if it accurately predict the one left training example or 0 if doesn't. Therefore the minimization cost function will have $1/52 = 0.0192$ step. Because SFS is computationally heavy operation, not all models are suitable for this technique, especially TBagger and ANN.

4.3   Voting from Ensemble Bucket of Models

After selecting suitable features for each model, we ensemble a model consisting of the five models. When we predict we would train all 5 models with the training data and predict with all of them using the test data. We get the consensus from at least 3 of the models to select the result.

5   Results for Inter-Subject Classification

5.1   Classification Using All Features

In Table 1 are given the prediction accuracy results using all features for test and train data. Comparison of the prediction accuracy using all features and the selected features is shown on Fig. 2.

Table 1. Prediction accuracy results from classification models using all features

Model	ANN	LogReg	LDA	kNN	NB	SVM	DT	Tbagger
Accuracy X$_{TEST}$	71,2	67,31	71,2	59,6	69,2	50	69,2	75
Accuracy X$_{TRAIN}$	75,6	100	100	100	93	100	96,2	100

5.2   PCA Feature Reduction and Classification

After calculating eigenvectors we estimate the numbers of vectors used to project the data with 99%, 95%, 75% and 50% data variance retained corresponding number of features is 43, 34, 16, and 7. Results from the prediction accuracies can be seen in Table 2. It is seen that we cannot improve significantly prediction accuracy using PCA and data projection in lower dimensionality.

Table 2. Results from models using reduced (projected) by PCA features set

Model	ANN	LogReg	LDA	kNN	NB	SVM	DT	Tbagger
43 Features (99%)	53,9	67,31	65,4	59,6	61,5	57,7	48,1	57,69
34 Features (95%)	61,5	69,23	65,4	57,7	67,3	57,7	53,9	63,46
16 Features (75%)	57,7	71,15	67,3	55,8	65,4	63,5	63,5	69,23
7 Features (50%)	55,8	59,62	61,5	69,2	65,4	71,2	63,5	61,54

5.3   Exhaustive Sequential Feature Selection (SFS) and Classification

Exhaustive SFS is computationally very intensive operation, therefore the SFS was performed on a smaller set of ML models. The resulting cost function (1-accuracy) based on the number of selected features is depicted on Fig. 1. Note that the number of features that minimizes the cost function is different for each model, typically between 5 and 10.

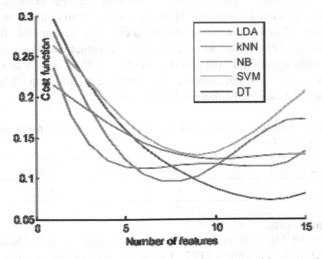

Fig. 1. Features selection: Cost function on numbers of features selected

Table 3. Features selected by SFS for each model

Features	LDA		kNN		NB		SVM		DT	
Number	Ch.	Feature	Ch.	Feature	Ch.	Feature	Ch.	Feature	Ch.	Feature
1	1	amp5	4	amp1	1	amp4	1	amp6	1	amp4
2	3	amp1	5	amp1	2	latency4	3	latency1	12	amp3
3	5	latency2	8	latency3	3	amp1	5	latency4	14	amp2
4	6	latency3	10	amp1	4	amp6	5	latency5	20	latency4
5	6	latency4	10	amp6	9	amp2	11	latency6		
6	11	amp3	13	amp3	20	latency4	13	amp2		
7	13	amp1	14	latency4						
8	13	amp6	20	amp2						
9	17	latency6								
10	20	latency4								

Table 4. Prediction accuracy on test and train data for models trained using the selected features from Table 2

Model	LDA	kNN	NB	SVM	DT
Accuracy X_{TEST}	92,3	90,38	86,5	88,5	88,5
Accuracy X_{TRAIN}	94,2	100	91,2	100	97,4

5.4   Voting from Ensemble Bucket of Models

The combination of SFS and the five ML classifiers in the previous section brought already results very close or even slightly better than the best classification rates published in previous related researches. However, we made an intuitive step ahead to build an ensemble classifier based on the majority vote among the five trained models. Thus, the prediction rates achieved by the individual classifiers in the range of [87 – 92] %. were significantly improved and achieved 98%, see Table 4.

4). Finally we can observe and compare the prediction accuracy on all features and selected features and ensemble bucket models vote in fig. 2.

Table 5. Accuracy and confusion matrix on test data using voting from models trained using the selected features from Table 2

Accuracy X_{TEST}	True 1	False 1	False 0	True 0
98,08	26	0	1	25

Discussion of the Results

We used supervised ML methods to predict two human emotions based on 252 features collected from 21 channels EEG. The achieved prediction accuracy based on all features is in the range of 60-75% (see Table 1). These results are similar to other related studies, [6] and they can be explained by the limited examples in the data set (2 examples per subject, 26 subjects, that corresponds to 52 examples in total) and the very high dimensional feature space (252). It was expected that predictions based on reduced number of features will perform better. While the PCA feature reduction did not bring any improvement (see Table 2), the Sequential Feature Selection (SFS) reduced the feature set to 4-10 features (see Fig. 1 and Table 2) and significantly improve the prediction accuracy of all studied ML models in the range of 88-92 % (Table 3). Finally, our empirical approach of combining the five previous classifiers in an ensemble bucket of models and use the majority vote as the final attributed class further improve substantially the prediction accuracy to 98% (Table 4). This is the main contribution of this paper, because such inter-subject classification accuracy was never before reported. The influence of the SFS is visualized on Fig. 2. We may also argue that our models can be used in real time, because after finding off-line the right features and training, the feature generation from monitored EEG signals is less than 1000ms and prediction is instantaneous.

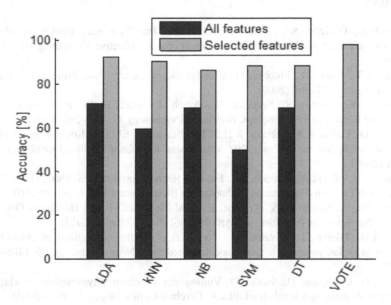

Fig. 2. Classification accuracy on test data. 5 classifiers (LDA, kNN, NB, SVM, DT) and their majority vote combination (VOTE).

6   Conclusion

In this paper, we have presented results demonstrating the feasibility of ML classification techniques to distinguish the processing of stimuli with positive and negative emotion valence based on ERPs observations. This problem is interesting both because of its relevance to studying human emotions, and as a case study of supervised machine learning (ML) in high dimensional data settings. The focus of our work was to explore the feasibility of training cross-subject classifiers to make predictions across multiple human subjects. Feature selection is an important aspect in the design of the recognition systems, particularly in the inter-subject framework. The combination of adequate features and channel selection has the potential to reduce the inter-subject variability and improve the learning of representative models valid across multiple subjects.

It can be concluded that ML is a powerful technique to reveal the brain activity and to interpret human emotions. There are many additional opportunities for ML research in the context of affective neuroscience, such as discrimination of more than two emotional states related not only with the emotional valence but also with the emotional arousal. Discrimination of high versus low neurotic type of personality is also a challenging problem that ML can deal.

Acknowledgements. We would like to express thanks to the PsyLab from Departamento de Educação da UA, and particularly to Dr. Isabel Santos, for providing the data sets.

References

1. Calvo, R.A., D'Mello, S.K.: Affect Detection: An Interdisciplinary Review of Models, Methods, and their Applications. IEEE Transactions on Affective Computing 1(1), 18–37 (2010)
2. Dalgleish, T., Dunn, B., Mobbs, D.: Affective Neuroscience: Past, Present, and Future. Emotion Rev. 1, 355–368 (2009)
3. Olofsson, J.K., Nordin, S., Sequeira, H., Polich, J.: Affective Picture Processing: An Integrative Review of ERP Findings. Biological Psychology 77, 247–265 (2008)
4. AlZoubi, O., Calvo, R.A., Stevens, R.H.: Classification of EEG for Emotion Recognition: An Adaptive Approach. In: Proc. 22nd Australasian Joint Conf. Artificial Intelligence, pp. 52–61 (2009)
5. Petrantonakis, P.C., Hadjileontiadis, L.J.: Emotion Recognition from EEC Using Higher Order Crossings. IEEE Trans. Information Technology in Biomedicine 14(2), 186–194 (2010)
6. Jatupaiboon, N., Panngum, S., Israsena, P.: Real-Time EEG-Based Happiness Detection System. The ScientificWorld Journal, Article ID 618649, 12 pages (2013)
7. Santos, I.M., Iglesias, J., Olivares, E.I., Young, A.W.: Differential effects of object-based attention on evoked potentials to fearful and disgusted faces. Neuropsychologia 46, 1468–1479 (2008)
8. Pourtois, G., Grandjean, D., Sander, D., Vuilleumier, P.: Electrophysiological correlates of rapid spatial orienting towards fearful faces. Cerebral Cortex 14(6), 619–633 (2004)
9. Lecture 13: Validation, http://research.cs.tamu.edu/prism/lectures/iss/iss_l13.pdf
10. CS229 Machine Learning, Andrew Ng, http://cs229.stanford.edu/
11. Bishop, C.M.: Pattern Recognition and Machine Learning. Springer (2006)
12. Fisher, R.A.: The Use of Multiple Measurements in Taxonomic Problems. Annals of Eugenics 7, 179–188 (1936)
13. Matlab documentation, http://www.mathworks.com/help/matlab/
14. Palaniappana, R., Ravi, K.V.R.: Improving visual evoked potential feature classification for person recognition using PCA and normalization (2005)

Application of Neural Networks Solar Radiation Prediction for Hybrid Renewable Energy Systems

P. Chatziagorakis[1], C. Elmasides[2], G.Ch. Sirakoulis[1,*], I. Karafyllidis[1], I. Andreadis[1], N. Georgoulas[1], D. Giaouris[3], A. I. Papadopoulos[3], C. Ziogou[3], D. Ipsakis[3], S. Papadopoulou[3], P. Seferlis[3], F. Stergiopoulos[3], and S. Voutetakis[3]

[1] Department of Electrical and Computer Engineering, Democritus University of Thrace, Xanthi, Greece
[2] Systems Sunlight S.A., Xanthi, Greece
[3] Chemical Process Engineering Research Institute, Centre for Research and Technology Hellas, Thermi-Thessaloniki, Greece
gsirak@ee.duth.gr

Abstract. In this paper a Recurrent Neural Network (RNN) for solar radiation prediction is proposed for the enhancement of the Power Management Strategies (PMSs) of Hybrid Renewable Energy Systems (HYRES). The presented RNN can offer both daily and hourly prediction concerning solar irradiation forecasting. As a result, the proposed model can be used to predict the Photovoltaic Systems output of the HYRES and provide valuable feedback for PMSs of the understudy autonomous system. To do so a flexible network based design of the HYRES is used and, moreover, applied to a specific system located on Olvio, near Xanthi, Greece, as part of SYSTEMS SUNLIGHT S.A. facilities. As a result, the RNN after training with meteorological data of the aforementioned area is applied to the specific HYRES and successfully manages to enhance and optimize its PMS based on the provided solar radiation prediction.

Keywords: Recurrent Neural Network, Solar Radiation, Power Management Strategy, Hybrid Renewable Energy System.

1 Introduction

In recent years, as a response to the continuously growing need for green energy, a new type of Renewable Energy Systems (RES) is becoming all the more popular [1]. That is the *Hybrid Renewable Energy Systems* (HYRES). These systems often combine a variety of different renewable technologies with some energy storing units in a single generating plant facility. This combination offers the advantage of exploiting different types of green energy without completely depending on the availability of a single one. Therefore, hybrid systems present a better balance in energy production than the conventional systems, which make use of a single

---

* Corresponding author.

V. Mladenov et al. (Eds.): EANN 2014, CCIS 459, pp. 133–144, 2014.
© Springer International Publishing Switzerland 2014

technology and tend to be more inconsistent. The utilization of multiple green energy sources provides to these systems increased efficiency, as well as balance in energy supply due to the fact that each energy source acts as supplement to the others. This is the reason why HYRES are considered as a reliable solution for remote area power generation applications.

However, despite the advantages that the adoption of HYRES may have, there are still some weak spots. Despite the fact that these systems rely on multiple renewable sources, they are still dependent on conventional fuels as long as the green energy is not always available. In addition, another characteristic that needs further improvement is their efficiency. Although they present a much higher efficiency than the single-technology RES, there is still space for further improvement and optimization [1]. The cooperation between the different discrete systems does not often occur in the most efficient way. For example, storage of the excess energy supply does not always occur in the most effective way and thus the system usually depends on conventional fuels.

The great dependency that HYRES efficiency has on both the availability and the values of the critical meteorological variables necessitates their a priori knowledge. Meteorological variables such as solar radiation, air temperature, wind speed and humidity can affect at a maximum degree the functionality and the efficiency of the corresponding RES. Having this kind of information someone can provide some satisfactory estimation of the total amount of the future renewable energy production. In this way, a better management of the HYRES subsystems can be achieved, as well as a much more optimized energy storage and utilization, thus diminishing the need for conventional generators. The optimized management of the various subsystems is the key point towards achieving the best possible green energy utilization and system efficiency.

The goal of the specific study is the design of an autonomous intelligent forecasting model based on Neural Networks (NNs) that will enable the future values estimation for the critical meteorological parameters, such as solar radiation, that greatly affect the efficiency and the overall functionality of the corresponding HYRES. During the previous years, NNs have been proposed as a powerful approach to estimate and predict the solar radiation in different areas all over the world [2-9]. Taking into account that the solar radiation time series as a dynamical system presents nonlinear characteristics due to its dependency on meteorological parameters, like temperature, water vapor, cloud and water air condition, etc. [10], we propose a Recurrent Neural Network (RNN) with Nonlinear Autoregressive architecture (NAR) as an enhanced forecasting model for solar radiation time series. The here proposed model will have the ability of assimilating the past meteorological datasets and thus learning the local behavior of the target parameters. For this reason, it will incorporate the feature of receiving the current meteorological data from locally installed sensing devices. Next, it will provide the corresponding future estimations through the combination of the past knowledge and current feedback.

Additionally, the presented model will also be applicable to an autonomous HYRES, where its estimations can be used by a central control unit in order to create in real time the proper Power Management Strategies (PMSs) for the efficient subsystems utilization that can lead to the overall optimization. For doing so, a generic network model is also described for the representation of the hybrid power generation systems taking into consideration in this work. Subsequently, the RNN when combined with the presented network model of HYRES serves as a novel framework for a generic approach aiming to facilitate the derivation of various PMSs in a simple and flexible way. As a result, the proposed framework will make the specific HYRES suitable for use as a standalone remote energy plant. As a proof of concept the results of the proposed NN model for solar radiation forecasting when applied to an available HYRES system are also presented. It is clear that the proposed RNN after training with meteorological data of the under study area, in our case Olvio of Xanthi in Greece and applied to the proposed HYRES of Systems Sunlight SA finally manages to enhance and optimize its PMS based on the provided solar radiation prediction.

2   The Proposed Recurrent Neural Network

Having in mind that the estimation of the future weather conditions constitutes a particularly complex problem due to the non-linear dynamics of the weather behavior, the computational paradigm of Neural Networks (NNs) was adopted for the design of the forecasting model. In specific, due to the nature of the problem that involves the estimation of the future values of certain meteorological variables, the adopted type was the *Recurrent Neural Network* (RNN) [11-14]. They constitute some NN type where the unit connections form some directed cycle. This inherent characteristic creates some internal network state that greatly favors the exhibition of some dynamic temporal behavior. Unlike conventional *Feed-Forward Neural Networks* (FF-NN), RNNs can use their internal memory to process arbitrary sequences of inputs.

As already mentioned, the RNN architecture differs from the FF-NN one to the point that except for the network inputs it also takes into consideration its internal state. This internal state can be considered as a trace of the previously presented network inputs that have already been processed. This feature provides to RNNs the ability of learning the temporal-sequential dependencies that may occur among the data of a time series. However, it is quite easy to understand the functionality of RNNs through a direct comparison with a simple FF-NN.

It is quite convenient to consider a simple layered architecture that consists of one input, one hidden and one output layer. Following, eq. (1) describes the mathematic relations that occur between the data of subsequent levels.

$$y_j(t) = f\left(net_j(t)\right), net_j(t) = \sum_{i}^{n} x_i(t) v_{ji} + \theta_j \tag{1}$$

where y is the layer output, j is the layer number, t is time, net_j describes the layer state and f is a differentiable output function. The variable n describes the total number of the network inputs x_i, whereas v_{ij} represent the connection weights and θ_j is a bias value. In case of a FF-NN the input array x is propagated through the weights V

that characterize the connection between the input and the hidden layer. Similarly, the propagation of the input array in a simple RNN is equally affected by the weights of the established connections between the nodes of the two neighboring layers. However, another factor that affects the propagation is an additional recurrent level that sends the previous network state through its own activation function and the corresponding connection weights U as presented by Fig. 1 and described by:

$$y_j(t) = f\left(net_j(t)\right), net_j(t) = \sum_i^n x_i(t)v_{ji} + \sum_h^m y_h(t-1)u_{jh} + \theta_j \tag{2}$$

where m expresses the total aggregate of the state nodes. The network output is defined in both cases by its own state and the weights W as follows:

$$y_k(t) = f\left(net_k(t)\right), net_k(t) = \sum_j^m x_j(t)v_{kj} + \theta_j \tag{3}$$

where g is the activation function of the output layer.

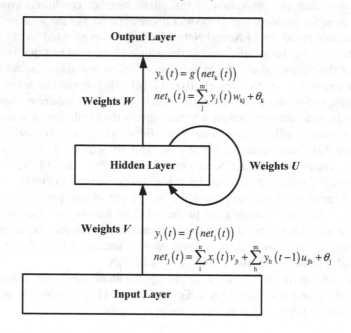

Fig. 1. Simplified Recurrent Artificial Neural Network (RANN) structure diagram

The selected architecture of RNN for the current study was the *Nonlinear Autoregressive Neural Network* (NAR). This NN has been extensively used for statistical forecasting modeling of time series [15-18]. The specific model constitutes dynamic RNN that includes feedback connections between layers. NAR is based on the *Linear Autoregressive Model* (AR), which is known for its effectiveness in modeling time series [18]. The equation that defines NAR functionality is:

$$y(t) = f\left(y(t-1), y(t-2), y(t-3), ..., y(t-k)\right) \tag{4}$$

where $y(t)$ is the model output that depends on the k temporally previous values of the output signal. This is also represented in Fig. 2, which shows the block diagram of a two-layer AR model with feed-forward architecture. The purpose of the specific model is the estimation of function f. The input of the model can be a multidimensional array, while each layer has an additional bias input b_1 and b_2 for quicker convergence of the NN. Furthermore, the connection between two layers is characterized by the corresponding layer weight $LW_{i,j}$, where i refers to current layer and j refers to previous layer. Finally, the *Time Delayed Line* (TDL) expresses the time delay that is inflicted upon the output feedback data that are sent back to the input through the feedback loop. This feature enables the estimation of the temporal dependencies that may occur between the input and the output. This property is very helpful when trying to model systems that are described by time series.

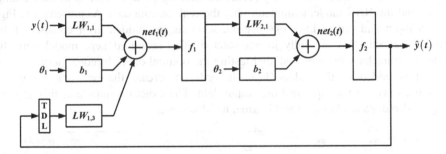

Fig. 2. Autoregressive Model (AR) architecture

In general, there are two different architecture options for NAR. Both these architectures include a time delay line that was described above. However, the first one includes a feedback loop that sends the data from the output directly back to the input (so named parallel) and the serial NAR architecture which lacks any feedback property. The main difference between these two options relates to the training procedure. The accuracy of the training is higher in the second case, because through the serial feed-forward architecture the network is fed only with real data. On the contrary, the parallel NAR combines both feedback and real data. This often has a negative effect at the network training accuracy as long as the output data are already processed. Another advantage of the serial model over the parallel one is its simplicity. The serial architecture produces more responsive models that are easier to implement and train faster. This is considered as a very significant feature when the NAR model is meant to be used in real time applications.

For the needs of the current study, the proposed NAR model was initially trained through the adoption of the serial architecture. The basic structure of the model has three distinct layers; the input, the hidden layer and the output. The utilization of a single hidden layer was decided upon the fact that in literature there are a lot of NN examples where such architecture provides enough computational power for confronting problems of similar complexity [2]-[8]. Moreover, in order to decide which NAR architecture and network size is the most suitable, a series of different tests were realized that confirm the suitability of the aforementioned structure and resulted to network efficiency. Making use of a small but representative dataset, a variety of networks was tested including NARs with multiple hidden layers and variable neurons numbers per layer.

The proposed NAR model was designed and simulated through the use of the *Neural Time Series Tool* that constitutes part of the *Neural Network Toolbox* library of *Matlab* software. The design of the NAR model has been done in accordance with the network architecture of Fig. 2. As mentioned before, a single hidden layer was used, as it is considered to deliver some adequate computational power and network efficiency in combination with some good performance [5], [7]. The NAR model as designed with Neural Time Series Tool is presented in Fig. 3. The specific architecture does not include any output feedback property so as to deliver the best possible training results. This open loop network was used for training purposes, as long as its output is not sent back to the input. In this way, every presented training pattern belongs to the training dataset and the NAR model assimilates only the real training data. As presented in Fig. 3, both input and output have one neuron, whereas the hidden layer consists of 10 artificial neurons that are fully interconnected. The proposed here model with 10 hidden neurons have been proved after testing the optimal one. Moreover, a time delay element is added to the hidden layer in order to create the necessary temporal difference between the input and the output data. The exact magnitude of this applied delay is also depicted in Fig. 3 and is equal to 72 samples.

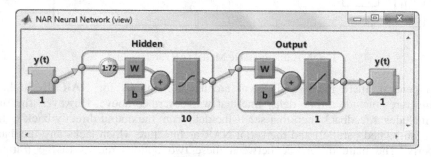

Fig. 3. Schematics of the open loop Nonlinear Autoregressive Network (NAR) designed with Neural Time Series Tool

Following the network design, the available real meteorological data were used for its training. In accordance to the experimental setup that will be introduced in Section 3, the training time series were collected from the location of Olvio with coordinates 41.0249 N and 24.7885 E near the city of Xanthi in Greece. These datasets include solar radiation measurements for the time period of the last two years, having a time resolution of 5 minutes. Some representative examples of solar radiation datasets are demonstrated in Fig. 4(a) – 4(b), in specific, for the months October and June in respect. It can be easily noticed that every month has its own characteristics, present some nonlinear behavior and can potentially present great differences from one day to the other. The differences between months are considered normal due to the different season conditions. Furthermore, some intense fluctuations that take place during the same month often are caused by the occurrence of some extreme weather phenomena. The time resolution of the measurements is equal to 1 sample per 5 minutes. This means that 288 samples are available per day. However, for the needs of the specific study three additional datasets with lower resolution were created and tested; with 96, 24 and 4 samples/day. Through the realization of different training scenarios for the four datasets, it was proved that 24 samples/day provided the best results in terms of accuracy and computational speed, without lacking any significant information comparing to the initial dataset. By adopting the 24 samples/day dataset with an input delay line of 72 samples, the proposed NAR model receives the measurements of the three last days as an input to estimate the future value.

Additionally, before feeding the designed model with the training dataset, two techniques were used in order to improve the learning efficiency. Namely with the help of Matlab, solar radiation during nighttime which was always equals to zero, thus disrupting the consistency of the training patterns, has been deleted and the variation of solar radiation during the sunlight hours is only considered. Moreover, normalization of synaptic weights updating process takes place so as prevent uncommonly high training values from affecting the overall NAR training accuracy. Afterwards, the learning dataset was divided into the necessary subsets; training, validation and testing. The adopted ratios were equal to 75%, 15% and 10% for the training, validation and testing subsets in respect after several tests.

Next, the *Levenberg - Marquardt* back-propagation algorithm was adopted for the training of the proposed NAR model. The specific technique is widely used for the NN training as it is considered as one of the most efficient solutions. During the learning procedure it implements the error back-propagation method, whereas it updates the synaptic weights and bias values according to Levenberg-Marquardt optimization [19]. There is a great variety of different approaches, such as *Bayesian Regulation Back-propagation*, *Gradient Descent Back-propagation* and *BFGS Quasi-Newton Back-propagation* [20]. Bearing that in mind, all the above learning algorithms were tested in training the proposed model. However, deliver the best training results in combination with the quickest convergence were achieved throught Levenberg-Marquardt algorithm.

The success of the network training does not only depend on the available dataset and the learning algorithm. Defining the basic training parameters and the ending conditions of the learning procedure is also a significant task. At first, the *Mean Squared Normalized Error* (MSE) performance function was adopted [21]. It measures the network performance according to the mean of squared errors, i.e. it calculates the produced error that is the distance between the NAR actual outputs and the desired target values. Next, the maximum duration of the NAR training was specified equal to 10^3, whereas the maximum time period was defined to be 60 minutes. Moreover, the minimum gradient limit was defined as equal to 10^{-5} and the maximum number of validation fails was equal to 6. The occurrence of successive validation fails often means that the designed model has reached the maximum possible assimilation level for the given training dataset and network architecture. Finally, the training results were quite satisfactory bearing in mind the complexity of the target system. In specific, the final MSE was equal to 70 W/m^2 that is considered quite decent regarding the actual solar radiation datasets values that may vary up to 1300 W/m^2.

Following, the trained NAR model was tested on real solar radiation data for the creation of next day estimations. The input datasets are real random data that were acquired from the same location as the training set, but were completely excluded from the network training procedure. The corresponding testing results can be found in Fig. 4(c)-4(d). The actual solar radiation values are represented by the blue line, whereas the predicted values by the red line. Samples 1-72 constitute the model input and samples 73-96 are the produced future estimations. For this reason, the most significant part of the diagram is the comparison between the two representations regarding during these last 24 samples. However, for demonstrative reasons the rest of the samples were included.

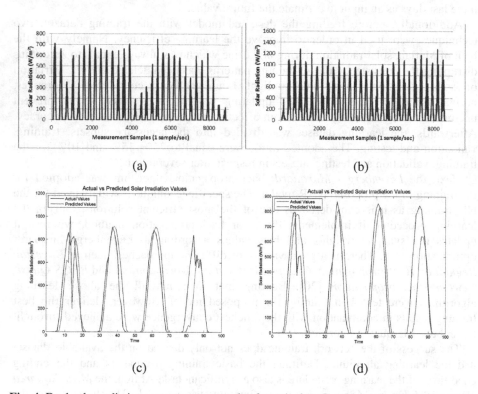

Fig. 4. Real solar radiation measurement samples from the location Olvio (41.0249 N, 24.7885 E), Xanthi, Greece for the months: (a) October and (b) June and (c), (d) comparison between the produced forecasting results of the proposed NAR model (red line) and corresponding real solar radiation measurements on the horizontal level (blue line) for two different time periods.

3   Efficient Representation of Energy Management Strategies

In this section we will review the representation of PMSs as described in [22] where the microgrid was seen as a graph and the flow of power and hydrogen within was described through flowsheets. The basic devices are the Photovoltaic Panels (PV), the Wind Generators (WGs), the Battery (BAT), the Fuel Cell (FC) and an Electrolyser (EL). In [22] each device in the microgrid is seen as a node of a graph and its connection as an edge. In our system the flows between the nodes can be in various states like electrical energy (POW) or hydrogen in high pressure (H2P) and hence the input to each node for each state j is given by

$$F_n^{In,j}(t) = SF_n^j(t) + \sum_{l=1}^{N} \varepsilon_{l \to n}(t) F_{l \to n}^{Out,j}(t)$$ where $F_n^{In,j}(t)$ is the input to node n at

the instant t, $SF_n^j(t)$ are external inputs, $F_{l \to n}^{Out,j}$ are the outputs of the other nodes, $\varepsilon_{l \to n}(t)$ are binary variables that determine the connection of a specific edge and N is the number of nodes in the graph.

The binary variables that determine the connection can be defined as:

$$\varepsilon_{l \to n}(t) = L\left(\varepsilon_{l \to n}^{Avl}(t), \varepsilon_{l \to n}^{Req}(t), \varepsilon_{l \to n}^{Gen}(t)\right) \qquad (5)$$

where L is a logical operator (like AND, OR, ...) and $\varepsilon_{l \to n}^{Avl}(t), \varepsilon_{l \to n}^{Req}(t), \varepsilon_{l \to n}^{Gen}(t)$ are binary variables that determine the availability, the requirement and other general conditions necessary to activate the connection l to n. In general the activation of a connection depends on logical propositions that can be described by binary variables ρ_i. Using this approach it is possible to systematically represent any PMS for a microgrid. For example another PMS can be one where the operation of the devices depend on the time of the year, i.e. during the summer the FC is not activated even if the SOC drops below $str_{FC \to BAT}^{SOAcc^{BAT}}$, similarly for the EL during the winter months.

Thus $\varepsilon_{FC \to BAT}(t)$ is written as:

$$\varepsilon_{FC \to BAT}^{Gen}(t) = 0 \quad t \in [2881, 5832] \qquad (6)$$

4   Combination of the Flexible PMS Representation and Solar Radiation Forecast

In order to test the efficiency of the proposed model an already implemented system HYRES was taken under consideration. As a result, the meteorological measurements used in this study are real data collected from the location of Olvio near the city of Xanthi in Greece, where the main compound of SYSTEMS SUNLIGHT S.A. is located. In Fig. 5 there is a block diagram describing the general architecture and components of the available HYRES system. In the following we use the solar radiation prediction RNN presented in Section 2 and we combine it with the PMS representation of Section 3. As it has been said, when the 2nd PMS is utilised, the FC is not allowed to operate during the summer months even in the case where the SOC is low and there is available hydrogen. The main argument for that is that there will be intense solar irradiation after a few hours and hence energy will be produced that not only will charge the battery. While in most cases this approach will reduce the usage of the FC without over depleting the battery in some cases it is possible to cause many problems. This is happening if during the next 24h there is a rather low solar irradiation and hence the DSL is forced to be activated. This issue gets more serious if there are multiple successive days with low solar irradiation. This problem can be overcome if using the weather forecast method explained before we detect the days with low solar irradiation and for these days we allow the usage of the FC (i.e. operate under the first PMS). As a case study in Figure 6(a) we see the power produced by the PVs in the system for 4 days during August. We see that on the 3rd of August the maximum power is above 12kW while in the 5th less than 8kW. More specifically the total energy produced during these 4 days is approximately 92kWh, 41kWh, 48kWh and 67kWh respectively. In this case study the load was fixed at 3kW and hence each day there is a requirement of 72kWh. The FC was operated at 2kW and the DSL at 3kW (in order to protect the battery).

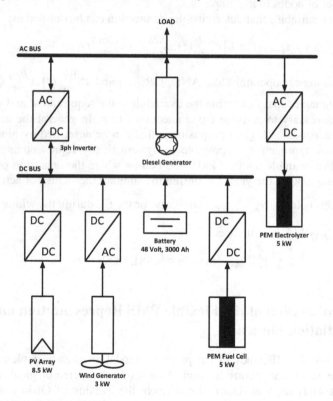

Fig. 5. Schematic of the available Hybrid Renewable Energy Systems (HYRES) of the Sunlight Systems S.A. that is installed in the location of Olvio, Xanthi, Greece and is used for the needs of the current research

In Fig. 6(b) we see the state of charge of the battery when the two PMS were used. In the first case (solid trace) the PMS did not allow the activation of the FC and while this did not cause any problems during day 1 (3rd of August) at the end of the second day the SOC dropped below 0.2 and the DSL was activated. On the other hand when based on the weather prediction, it was seen by the system that on the 4th and 5th of August the solar irradiation may not be high enough, the PMS was changed and the FC was activated which stopped the SOC dropping below 0.26.

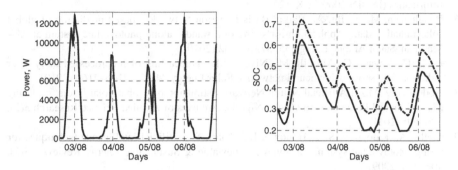

Fig. 6. (a) Power produced by the PVs. (b) SOC response under the 2 PMSs

5   Conclusions

In this paper a RNN for solar radiation prediction is proposed for the enhancement of the PMSs of HYRES. The presented RNN with NAR architecture can offer both daily and hourly prediction concerning solar irradiation forecasting. As a result, the proposed model can be used to predict the Photovoltaic Systems output of the HYRES and provide valuable feedback for PMSs of the understudy autonomous system. To do so a flexible network based design of the HYRES is used and, moreover, applied to a specific system located on Olvio, near Xanthi, Greece part of SYSTEMS SUNLIGHT facilities. As a result, the proposed RNN was trained with meteorological data of the aforementioned area and then applied to the proposed HYRES managing to enhance and optimize its PMS based on the provided solar radiation prediction. Furthermore, as a future work we will apply the presented RNN to estimate wind speed aiming at the optimal utilization of both solar and wind energy through a combination of autonomous hybrid systems that gather multiple renewable sources as already proposed.

Acknowledgements. This work is co-financed by National Strategic Reference Framework (NSRF) 2007-2013 of Greece and the European Union, research program "SYNERGASIA" (SUPERMICRO – 09ΣYN-32-594).

References

1. Deshmukha, M.K., Deshmukh, S.S.: Modeling of hybrid renewable energy systems. Renewable and Sustainable Energy Reviews 12(1), 235–249 (2008)
2. Alam, S., Kaushik, S.C., Garg, S.N.: Computation of beam solar radiation at normal incidence using artificial neural network. Renewable Energy 31(10), 1483–1491 (2006)
3. Mubiru, J., Banda, E.J.K.B.: Estimation of monthly average daily global solar irradiation using artificial neural networks. Solar Energy 82(2), 181–187 (2008)

4. Rehman, S., Mohandes, M.: Artificial neural network estimation of global solar radiation using air temperature and relative humidity. Energy Policy 36(2), 571–576 (2008)
5. Ghanbarzadeh, A., Noghrehabadi, R., Assareh, E., Behrang, M.A.: Solar radiation forecasting using meteorological data. In: 7th IEEE International Conference on Industrial Informatics (INDIN 2009), UK, (2009)
6. Benghanem, M., Mellit, A.: Radial Basis Function Network – based prediction of global solar radiation data: Application for sizing of a stand – alone photovoltaic system at Al – Madinah, Saudi Arabia. Energy 35, 3751–3762 (2010)
7. Paoli, C., Voyant, C., Muselli, M., Nivet, M.L.: Forecasting of preprocessed daily solar radiation time series using neural networks. Solar Energy 84(12), 2146–2160 (2010)
8. AbdulAzeez, M.A.: Artificial Neural Network Estimation of Global Solar Radiation Using Meteorological Parameters in Gusau, Nigeria. Archives of Applied Science Research 3(2), 586–595 (2011)
9. Mellit, A., Kalogirou, S.A., Hontoria, L., Shaari, S.: Artificial intelligence techniques for sizing photovoltaic systems: a review. Renewable & Sustainable Energy Reviews 13(2), 406–419 (2009)
10. Zeng, Z., Yang, H., Zhao, R., Meng, J.: Nonlinear characteristics of observed solar radiation data. Solar Energy 87, 204–218 (2013)
11. Grossberg, S.: Nonlinear neural networks: Principles, mechanisms, and architectures. Neural Networks 1, 17–61 (1988)
12. Anderson, J.A.: Introduction to Neural Networks. MIT Press, Cambridge (1995)
13. Elman, J.: Finding structure in time. Cognitive Sci. 14, 179–211 (1990)
14. Pearlmutter, B.A.: Gradient calculations for dynamic recurrent neural networks: a survey. IEEE Transactions on Neural Networks 6(5), 1212–1228 (1995)
15. Hwang, S.Y., Basawa, I.V.: Large sample inference based on multiple observations from nonlinear autoregressive processes. Stochastic Processes and their Applications 49(1), 127–140 (1994)
16. Kapetanios, G.: Nonlinear autoregressive models and long memory. Economics Letters 91(3), 360–368 (2006)
17. Taskaya-Temizel, T., Casey, M.: A comparative study of autoregressive neural network hybrids. Neural Networks 18(5-6), 781–789 (2005)
18. Guo, W.W., Xue, H.: Crop Yield Forecasting Using Artificial Neural Networks: A Comparison between Spatial and Temporal Models. Mathematical Problems in Engineering 857865, 7 (2014)
19. Kohonen, T.: Self – Organization and Associative Memory. Springer (1989)
20. Haykin, S.: Neural Networks: A Comprehensive Foundation. Prentice Hall (1998)
21. Anderson, J.A., Rosenfield, E.: Neurocomputing: Foundations of Research. MIT Press (1989)
22. Giaouris, D., Papadopoulos, A.I., Ziogou, C., Ipsakis, D., Voutetakis, S., Papadopoulou, S., Seferlis, P., Stergiopoulos, F., Elmasides, C.: Performance investigation of a hybrid renewable power generation and storage system using systemic power management models. Energy 61, 621–635 (2013)

A New User Similarity Computation Method for Collaborative Filtering Using Artificial Neural Network

Noman Bin Mannan[1], Sheikh Muhammad Sarwar[1], and Najeeb Elahi[2]

[1] Institute of Information Technology, University of Dhaka,
Dhaka, Bangladesh
nomanbinmannan@gmail.com, smsarwar@du.ac.bd
[2] UiT The Arctic University of Norway
najeeb.elahi@uit.no

Abstract. A User-User Collaborative Filtering (CF) algorithm predicts the rating of a particular item for a given user based on the judgment of other users, who are similar to the given user. Hence, measuring similarity between two users turns out to be a crucial and challenging task as the similarity function is the core component of the item rating prediction function for a particular user. In this paper, we investigate the effectiveness of a multilayer feed-forward artificial neural network as a similarity measurement function. We model similarity between two users as a function that consists of a set of adaptive weights and attempt to train a neural network to optimize the weights. Specifically, our contribution lies in designing an error function for the neural network, which optimizes the network and sets weights in such a way that enables the neural network to produce a reasonable similarity value between two users as its output. Through experimentation on Movielens dataset, we conclude that neural network, as a similarity function, gains more accuracy and coverage compared to the Genetic Algorithm (GA) based similarity architecture proposed by Bobadilla et al.

Keywords: Collaborative filtering, Recommender System, Similarity measures, Artificial Neural Network.

1 Introduction

Recommender system (RS) makes custom-made recommendations to its users for products, services or information by applying various knowledge discovery techniques. The general objective of a RS is to predict rating of items of which the user has no knowledge. Different filtering algorithms are used in RS and a filtering algorithm is the core component of a recommender system. Most common filtering algorithms are demographic filtering [1] and content based filtering [2]. Demographic filtering is established using an intuition that users with common personal attributes like sex, age, region etc. will also have common personal preferences. Content based filtering recommends items to the user according to

V. Mladenov et al. (Eds.): EANN 2014, CCIS 459, pp. 145–154, 2014.

the content of the previously preferred items. This algorithm analyzes user's past behavior and recommends items according to user's preference history.

In recent years, collaborative filtering (CF) has been the mostly used filtering algorithm [8] [9] in RS. This filtering algorithm is based on the assumption that rating prediction for an unknown item for a given user should be influenced by a neighborhood of users with which the given user is similar. The neighborhood of similar users usually rate an important number of items in a similar way as the given user. For example "A Beautiful Mind" movie could be highly rated for an individual based on the positive ratings of a group of similar people about this movie who also rated this movie very highly. This recommendation will often provide the user of the service with inspiring positive information from the collective knowledge of all other users of the service.

The importance of recommender systems is increasing day by day. In recent years, RS have played an important role in reducing the unnecessary information overload on those websites, where users have the option of voting for their preferences on a series of products or services. Most well-known example of RS will be movie recommendation websites (i.e. IMDb) not only in aspect to users but also for researchers [3]. There are also many other application fields of RS such as e-commerce [4], e-learning [5], digital libraries [6] and so on. RS have great rule in future and it's importance is increasing day by day.

Artificial neural network (ANN) is a machine learning tool, which can be used for generating a series of nonlinear result values for real-valued and vector-valued functions over continuous and discrete-valued attributes. ANNs are also strong to noise in the training data [7]. The contribution of this paper is the introduction and applicability of ANN as a new similarity function for collaborative filtering. We train the neural network to enable it to produce a similarity value between two users. Through experimentation we show that ANN performs well than Genetic Algorithm [13] in terms of Mean Absolute Error (MAE) and Coverage.

2   Background and Related Work

Similarity computation between users is the main task in collaborative filtering algorithms. For a User-User CF algorithm, similarity, $sim_{x,y}$, between the users x and y who have both rated the same items is calculated first. To calculate this similarity different metrics [10] [11] [12] are used.

Generally, for computing correlation-based similarity, similarity $sim_{x,y}$ between two users x and y is calculated using Pearson correlation or other correlation-based similarities. Other correlation-based similarities are constrained Pearson correlation, Spearman rank correlation and Kendalls τ correlation. In constrained Pearson correlation, mid point is used in place of mean rate which is the main difference with Pearson correlation. Spearman rank correlation is similar to Pearson correlation, except that the ratings are ranks. Kendalls τ correlation is similar to the Spearman rank correlation, but instead of using ranks themselves, only the relative ranks are used to calculate the correlation [17] [18]. Vector-cosine based similarity metric use user as a vector of ratings and measure the rating vectors cosign angle [15]. There exists other useful similarity measures based on conditional probability [19] [20]. The goal of all the similarity measures is to produce appropriate similarity values between two users depending on their item rating vector.

2.1 Predicted Rating Computation and Mean Absolute Error (MAE)

To calculate the predicted rating p_x^i for user x of an item i, the following Deviation From Mean (DFM) as aggregation approach is used:

$$p_x^i = \bar{r}_x + \frac{\sum_{n\epsilon k_x}[sim_w(x,n) \times (r_n^i - \bar{r}_n)]}{\sum_{n\epsilon k_x} sim_w(x,n)} \tag{1}$$

Where $\bar{r}_x$ is the average of ratings made by the given user x and $\bar{r}_n$, r_n^i is the average of ratings and rating of the neighbor for that item respectively made by the neighbor n . After calculating every possible prediction according to the similarity function sim_w, the mean absolute error (MAE) of the RS is measured as following:

$$MAE = \frac{1}{U} \sum_{u\epsilon U} \frac{\sum_{i\epsilon I_u} |p_u^i - r_u^i|}{l_u} \tag{2}$$

When running the similarity function, U and I_u represent respectively the number of training users and the number of training items rated by the user u.

2.2 Similarity Method Using Genetic Algorithm(GA Method)

The main goal of a CF based RS is to obtain better rating prediction for an unknown item by applying a similarity function that improves the accuracy [8] [11] of prediction of CF based RS. For this purpose, Bobadilla et al. proposed a genetic algorithm based similarity method.

First they generate some vector values of a user subject to another user for obtaining similarity between each pair of users. Then vector values are passed to a similarity function which is associated with some weight vectors. Weight vectors for optimal similarity function are obtained by genetic algorithm.

Genetic algorithm (GA) based similarity metric [13] is described below:

Vector Values. The pre-processing stage of the GA based method involves the computation of a vector between two users. Later, the vector is used to asses the similarity between the users. In order to understand the vector computation let us consider a RS with a set of U users, $(1, ..., U)$, and a set of I items $(1, ..., I)$. Users rate those items with a discrete range of possible values $(m, ..., M)$, where value m represents a scenario where the user is completely unsatisfied and value M indicates a situation where the user is completely satisfied.

The ratings made by a particular user x can be represented by a list, $r_x = r_x^{(1)}, r_x^{(2)}, ..., r_x^{(l)}$, where I is the number of items in the RS and r_x^i represents the rating that the user x has made over the item i. If an item is not rated by the user, mark $\bullet$ is used and therefore the expression $r_x^i = \bullet$ states that the user x has not rated the item i yet.

To compare both user x and y, their rating lists, r_x, r_y are compiled to another vector:

$$v_{x,y} = (v_{x,y}^{(0)}, ..., v_{x,y}^{(M-m)}) \tag{3}$$

whose dimension is the number of the possible ratings that a user can make over an item. Each component $v_{x,y}^{(i)}$ of the vector $v_{x,y}$, represents the ratio of items, j, rated by both users (that is to say, $r_x^{(j)} \neq \bullet$ and $r_y^{(j)} \neq \bullet$) and over which the absolute difference between the ratings of both users is $i (|r_x^i - r_y^i| = i)$, to the number of items rated by both users. That is to say, $v_{x,y}^{(i)} = a/b$ where b is the number of items rated by both users, and a is the number of items rated by both users over which the absolute difference in the ratings of both users is i.

In such a way the vector $v_{x,y}$ is produced from the rating lists of user x and user y.

Similarity Function for GA Based Architecture. The resultant vector produced from two users is used to compute similarity between two users. For similarity calculation between two users using the vector, GA method considers the following equation:

$$sim_w(x,y) = \frac{1}{M-m+1} \sum_{i=0}^{M-m} w^{(i)} v_{x,y}^{(i)} \tag{4}$$

Here, GA method introduces a weight vector $w = w^{(0)}, ..., w^{(M-m)}$, whose components lie in the range $[-1, 1]$ (that is to say, $w^{(i)} \in [-1, 1]$). The rationale for assigning different weights to different components is to indicate relative importance of different components. As the first scalar component of the vector denotes the number of movies on which two users completely agree with each other, it would possibly have higher value that other scalar components.

Genetic Algorithm (GA) Method. In order to find an optimal similarity function, sim_w, genetic algorithm has been used to search for an optimal weight vector w, which is associated with the optimal similarity function sim_w (Eq. 4). In this context, genetic algorithm performs a supervised learning task [14], whose fitness function or evaluation function is the Mean Absolute Error (MAE) of the RS. The population of the genetic algorithm is the set of different vectors of weights, w. This method stops to generate population when the output of the population evaluation function (MAE) of the RS is lower then a threshold value, γ.

3   New Similarity Computation Method Using ANN

In this section we present our rationale for using ANN as a similarity function. Moreover, we show that how we have modeled the similarity function using an ANN. Specifically, we design the objective function or cost function in such a way that the ANN being trained with substantial amount of instances produce satisfactory similarity values.

3.1   Rationale Behind Choosing Neural Network

In [13], genetic algorithm has been used to find optimal weight vector for the equation in 4. Genetic algorithms perform well with chromosomes represented as binary string, and even though there are methods in existing literature to represent chromosome as a floating point vector, genetic algorithms sometimes perform poorly when used for floating point weight adjustment. Moreover, performance of a genetic algorithm is mostly dependent on the set of initial population, which if not carefully chosen according to a good heuristic can lead to non-optimal solutions. In this specific problem, a five-length floating point weight vector is found with genetic algorithm, which produces minimum MAE. A short note on MAE is given in section 2.1. We argue that we can have a large set of weight vectors to train if we use neural network, and we can make the weight vector as large as we can by increasing number of hidden layers and nodes in each hidden layer. Furthermore, ANN is naturally designed to handled floating point values and hence is perfectly suitable for finding acceptable similarity value between two users, if provided with vector values 2.2 between two users.

3.2   Modeling Similarity Function as ANN

As a similarity function, we have used a multilayer feed forward neural network with one input layer, one hidden layer and one output layer. The neural network is depicted in section 3.3 (Fig. 1). Vector value between two users is calculated using equation 3, and the resulting vector is modeled as the input of the network. In order to model similarity using ANN, we have used five input nodes in input layer and three hidden nodes in hidden layer. We obtain the expected similarity (Eq.7) from the output node and execute error back propagation algorithm until the expected similarity (minimum error function is set) is obtained from the output node. After training all users for all training item we expect to find an optimal neural network which, if give vector values between two users can produce satisfactory similarity value.

3.3   Neural Network Cost Function

The design of our cost function emerges from equation 1, which is used for predicting rating for an item in collaborative filtering for a given user based his similarity with other users. If we consider that rating for a given user (u_x) is

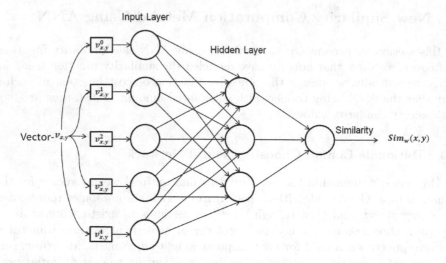

Fig. 1. Design of the similarity function as a multilayer feed-froward neural network

predicted using his similarity with another single user (u_n), the equation takes the following form:

$$p_x^i = \bar{r_x} + sim_w(x, n) \times (r_n^i - \bar{r_n}) \tag{5}$$

In the above equation, $\bar{r_x}$ is the average rating of u_x and $\bar{r_n}$ is the average rating of u_n. p_x^i is the predicted rating for u_x for item i, while r_x^i is the original rating of u_n for item i. Now, if we know the original rating r_x^i for u_x, we can write the above equation as below by slightly modifying it:

$$r_x^i = \bar{r_x} + sim_w(x, n) \times (r_n^i - \bar{r_n}) \tag{6}$$

From the above equation we can find the similarity as below:

$$sim_w(x, n) = \frac{r_x^i - \bar{r_x}}{r_n^i - \bar{r_n}} \tag{7}$$

So, if rating vectors of r_x and r_n are available we can always asses the similarity for each rating pair (r_x^i, r_n^i). The similarity $sim_w(x, n)$ would lie in the range of $[-1, 1]$.

For example, let us consider a rating database shown in table 1 for set of 6 users 1,2,..,6 and 6 items 1,2,..,6. Items are represented in the columns and users are shown in the rows. From the table we can obtain the rating vector for each of the users. Now, if we want to asses the similarity between the first user (u_1) and the third user (u_3) for movie 6, then the result will be as following:

$$sim_w(1, 3) = \frac{4 - 3}{3 - 2.33} = 1.49 \cong 1 \tag{8}$$

Table 1. Rating Table

	1	2	3	4	5	6	Avg. Rating
1	4	5	0	3	2	4	3
2	1	0	4	0	5	2	2
3	2	0	5	0	4	3	2.33
4	4	1	1	3	4	5	3
5	2	0	3	5	0	0	1.67
6	5	1	0	4	2	0	2

According to our design, we model ground truth similarity for neural network between two users for each item. By careful examination, it can be seen that the ground truth similarity between two users will change for each item. So, for each item if they have similar preference the neural network will have to go through less correction. So if, we have the rating vector for two users we can compute the vector values as shown in section 2.2, use them as input values for neural network and train the network to produce correct similarity value for each of the commonly rated items in the rating vector.

4   Experimental Results

4.1   Procedure

Data. We have run experiments with data from the MovieLens database developed by GroupLens and Internet Movie Database (IMDb). This database contain rating values for a set of movies by a set of users. We selected the first 1000 users as collaborative users that had rated more than 40 movies. The target users were selected from the users who's id was over 1000 (so that the collaborative group and the test group of users are disjoint) and had also rated approximately 30 movies.

Table 2. Descriptive information of the database used in the experiments

Dataset	Movielens
Users	6040
Movies	3952
Ratings	1000209
Min and Max Rating	1-5

Parameter Settings for Traning and Testing. For the training we have used 3952 collaborative users from the Movielens database. We train all users (3952) for a set of movies. After that we ran several experiments working with different parameters to find the best combination. After a set of experiments we obtained better result for a neural network with three hidden layers. The maximum number of epochs to train the network was five hundred and cross validation was used to test the neural networks performance.

Prediction and Recommendation Result. In this section, we show the results obtained using the dataset specified in Table. 2. The results of our proposed ANN method are compared with the ones obtained using GA method and traditional metrics on RS collaborative filtering: Pearson correlation, Cosine and Mean Squared Differences. The comparative results are shown in Fig. 2, in terms of Mean Absolute Error (MAE) and in terms of coverage they are shown in Fig. 3.

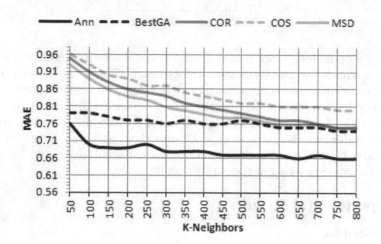

Fig. 2. Comparative resutls for GA, traditional metrics and proposed ANNs similarity method on Movielens dataset in terms of Mean Absolute Error

Fig 2 informs about the MAE obtained for Movielens when applying Pearson correlation (COR), cosine (COS), Mean Squared Differences (MSD), GA-method and the proposed ANNs method. The ANNs leads to fewer errors, particularly in the most used values of K. The black dashed and continuous lines represent respectively the best GA method and ANNs result.

Fig 3 informs about the coverage obtained. As may be seen, ANN method can improve the coverage for any value of K (the number of neighbors for each user) in relation to GA method and other traditional metric used.

The constant K is related to the number of neighbors for each given user and it varies between 50 and 800. These values enable us to view the trends of the graphics obtained from our ANN method compared to GA method and other traditional metric.

Graphic 2 shows that the best results in MAE with the ANN method are obtained when using a medium value in K, Graphic 3 shows that the best results with the ANN method are obtained in coverage using medium values in K. In this way, we should use intermediate values in K ($K \in \{300, ..., 400\}$) for obtaining the most satisfactory results both in MAE and in coverage. As our method provides high values in the quality measures applied on the MovieLens (mostly used database in RS) database, we can conclude that the proposed metric will work on a variety of Recommender Systems.

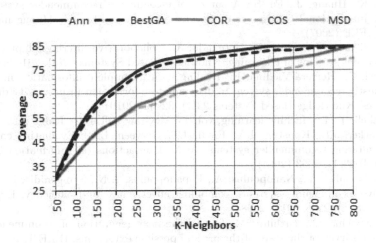

Fig. 3. Comparative resutls for GA, traditional metrics and proposed ANNs similarity method on Movielens dataset in terms of Coverage

5   Conclusion

The main contribution of this paper is the creation of a new similarity computation method using artificial neural network. When compared to GA method and other architectures, our similarity method architecture was able to reduce mean absolute error (MAE) convincingly. To be able to further evaluate our work it would be interesting to use singular value decomposition over our data. This would greatly reduce the amount of missing values in our dataset and most likely further increase the accuracy and coverage of our results. The MovieLens database have been extensively used in recommender systems research. Nevertheless, it would be useful to test the ANN based architecture with other datasets like eachmovie, film affinity, netflix and also other domains. Finally, even though applying neural network would slow down the process of recommendation generation by a few milliseconds if compared with the genetic algorithm (which gives results using a linear combination of weights), the MAE would be much less than the genetic algorithm based approach. Our experimental result clearly depicts this fact.

References

1. Krulwich, B.: Lifestyle finder: Intelligent user profiling using large-scale demographic data. AI Magazine 18(2), 37 (1997)
2. Lang, K.: Newsweeder: Learning to filter netnews. In: Proceedings of the Twelfth International Conference on Machine Learning, Citeseer (1995)
3. Miller, B.N., Konstan, J.A., Riedl, J.: Pocketlens: Toward a personal recommender system. ACM Transactions on Information Systems (TOIS) 22(3), 437–476 (2004)

4. Wei, K., Huang, J., Fu, S.: A survey of e-commerce recommender systems. In: 2007 International Conference on Service Systems and Service Management, pp. 1–5. IEEE (2007)
5. Bobadilla, J., Serradilla, F., Hernando, A.: Collaborative filtering adapted to recommender systems of e-learning. Knowledge-Based Systems 22(4), 261–265 (2009)
6. Porcel, C., Herrera-Viedma, E.: Dealing with incomplete information in a fuzzy linguistic recommender system to disseminate information in university digital libraries. Knowledge-Based Systems 23(1), 32–39 (2010)
7. Mitchell, T.M.: Machine learning, vol. 45. McGraw Hill, Burr Ridge (1997)
8. Herlocker, J.L., Konstan, J.A., Riedl, J.T., Terveen, L.G.: Evaluating collaborative filtering recommender systems. ACM Transactions on Information Systems (TOIS) 22(1), 5–53 (2004)
9. Manolopoulus, Y., Nanopoulus, A., Papadopoulus, A.N., Symeonidis, P.: Collaborative recommender systems: combining effectiveness and efficiency. Exp. Syst. Appl. 34(4), 2995–3013 (2008)
10. Adomavicius, G., Tuzhilin, A.: Toward the next generation of recommender systems: a survey of the state-of-the-art and possible extensions. IEEE Trans. Knowl. Data Eng. 17(6), 734–749 (2005)
11. Bobadilla, J., Serradilla, F., Bernal, J.: A new collaborative filtering metric that improves the behavior of recommender systems. Knowl. Based Syst. 23(6), 520–528 (2010)
12. Ingoo, H., Kyong, J.O., Tae, H.R.: The collaborative filtering recommendation based on SOM cluster-indexing CBR. Exp. Syst. Appl. 25, 413–423 (2003)
13. Bobadilla, J., Ortega, F., Hernando, A., Alcalá, J.: Improving collaborative filtering recommender system results and performance using genetic algorithms. Knowl. Based Syst. 24(8), 1310–1316 (2011)
14. Goldberg, D.E.: Genetic algorithms in search, optimization, and machine learning (1989) ISBN: 0-201-15767-5
15. Salton, G., McGill, M.: Introduction to Modern Information Retrieval. McGraw-Hill, New York (1983)
16. Resnick, P., Iacovou, N., Suchak, M., Bergstrom, P., Riedl, J.: Grouplens: an open architecture for collaborative filtering of netnews. In: Proceedings of the ACM Conference on Computer Supported Cooperative Work, New York, NY, USA, pp. 175–186 (1994)
17. Goldberg, K., Roeder, T., Gupta, D., Perkins, C.: Eigentaste: a constant time collaborative filtering algorithm. Information Retrieval 4(2), 133–151 (2001)
18. Herlocker, J.L., Konstan, J.A., Terveen, L.G., Riedl, J.T.: Evaluating collaborative filtering recommender systems. ACM Transactions on Information Systems 22(1), 5–53 (2004)
19. Karypis, G.: Evaluation of item-based top-N recommendation algorithms. In: Proceedings of the International Conference on Information and Knowledge Management (CIKM 2001), Atlanta, Ga, USA, pp. 247–254 (November 2001)
20. Deshpande, M., Karypis, G.: Item-based top-N recommendation algorithms. ACM Transactions on Information Systems 22(1), 143–177 (2004)

Probabilistic Models Based Intrusion Detection Using Sequence Characteristics in Control System Communication

Takashi Onoda

System Engineering System Laboratory,
Central Research Institute of Electric Power Industry,
2-11-1, Iwado Kita, Komae-shi, Tokyo 201-8511 Japan
onoda@criepi.denken.or.jp
http://www.criepi.denken.or.jp/en/index.html

Abstract. The importance of cyber security has increased with the networked and highly complex structure of computer systems, and the increased value of information. In this paper, we compare Conditional Random Field based intrusion detection with the other probabilistic models based intrusion detection. Theses methods uses the sequence characteristics of network traffic in the control system communication. The learning only utilizes normal data, assuming that there is no prior knowledge on attacks in the system. We applied these two probabilistic models to intrusion detection in DARPA data and an experimental control system network, and compared the differences in the performance.

Keywords: CRF, HMM, Control System Communication, Intrusion Detection, Sequence.

1 Introduction

The importance of cyber security has increased with the networked and highly complex structure of computer systems, and the increased value of information. Securing the network perimeter of a system is no longer considered sufficient to secure the system, and many layers of security, or defense in depth[1] is necessary.

Intrusion detection is one layer of security for a system, and is defined as "the act of detecting actions that attempt to compromise the confidentiality, integrity or availability of a resource"[2]. A basic intrusion detection in general checks IP (Internet Protocol) addresses, communication ports, and protocols at the TCP/UDP (Transmission Control Protocol/ User Datagram Protocol) level for non-legitimate connections. Constructing intrusion detection rules to match attacks does not necessarily require knowledge of the protocol details of the applications being used. Such rules are necessary to detect many attacks, and are proved to be highly effective [3,4].

We believe that it is effective to consider the sequence characteristics of the network traffic in control system communications. Previously, we investigated the use of intrusion detection systems in control systems that considered the application level behavior of the system[5,6]. We manually constructed intrusion detection rules checking

V. Mladenov et al. (Eds.): EANN 2014, CCIS 459, pp. 155–164, 2014.

the sequence characteristics of the network traffic caused by the control system protocol used by the control system application. The intrusion detection rules checked the traffic for sequence patterns that should not be possible during normal operation through the interface of that application.

Manually creating all the intrusion detection rules becomes impractical in a large system, and there has been work using machine learning for the creation of detection rules[7]. If there have been almost no observed cyber attacks, and very little attack data is available, it is difficult to construct a binary class classification problem for machine learning, which uses both the legitimate communication data and the attack data. In such cases, anomaly detection for intrusion detection would be a good choice[8]. There are many algorithms that can reflect sequence characteristics. The sequence data of the network traffic caused by the control system are probabilistic. In such a case, a Hidden Markov Model(HMM) and a Conditional Random Field(CRF) can reflect sequence characteristics. In this paper, we compare these two probabilistic models that reflect the sequence characteristics of network traffic into the intrusion detection system. The learning only utilizes normal data, assuming that there is no prior knowledge on attacks in the system. We show the results of the experiments.

2   Intrusion Detection Considering Sequences

Intrusion detection systems used in general computer systems widely use signature based systems, which build a model based on available knowledge of attacks. This has proved to be effective in a general computer system environment. But for new and unknown cyber attacks, there is no data that can be used to build a model for those attacks. It is relatively easy to collect a large amount of normal communication data; therefore, anomaly detection using only the available normal data becomes effective. However, signature based systems are superior when detecting known attacks, so either both types of intrusion detection would be applied, or we assume that the basic attacks would be prevented with the other security measures.

Another feature characteristic to control systems is the control system sequence in the network traffic. In a typical control system, many of the control system devices would be controlled in a certain order. For example, a certain valve would have to be closed before another valve can be opened. This is in order to prevent any physical damage to the control system. Generally, a violation of such a sequence would be prevented at the operator console, for example by not allowing the execution at the user interface, thus preventing any mistakes by the operator. There are usually also physical locks in the control system to prevent damage to the devices. However, if an attacker does not use the legitimate user interface and also eludes or disables the physical locks, there is a possibility that the attacker could cause damage to the control system. This would clearly violate the control system sequence, which should be visible in the network traffic, so if the violation could be detected by the intrusion detection system, it would help prevent damage to the system. Also, if the attacker does not have enough prior knowledge of the control system, the attacker may unknowingly violate the control system sequence. If the intrusion detection system is checking the sequences, such an attack would easily be detected.

With the features mentioned above, each sequence in the network traffic data is counted. Note that for the training, only the legitimate network traffic data of the system is used, making it adequate for control systems. After the training data is processed, the test data is processed with the same method, and if the data falls within a certain percentage of the minority data, the intrusion detection system alerts the data as an attack.

3   Hidden Markov Model and Conditional Random Field

In this section, we briefly introduce the two probabilistic models we used for intrusion detection; HMM and CRF.

3.1   HMM

The formal definition of a HMM is as follows:

$$\lambda = (A, B, \pi) \tag{1}$$

$S = (s_1, s_2, \cdots, s_N)$ is our state set, and $Y = (y_1, y_2, \cdots, y_N)$ is the observation set.

We define $Q = q_1, q_2, \cdots, q_T$, $O = o_1, o_2, \cdots, o_T$ to be a fixed state sequence of length T, and corresponding observations O.

$A = [a_{ij}]$, $a_{ij} = P(q_t = s_j | q_{t-1} = s_i)$ is a transition array, storing the probability of state j following state i. Note the state transition probabilities are independent of time.

$B = [b_i(k)]$, $b_i(k) = P(o_t = v_k | q_t = s_i)$ is the observation array, storing the probability of observation k being produced from the state j, independent of t.

$\pi = [\pi_i]$, $\pi = P(q_1 = s_i)$ is the initial probability array.

Two assumptions are made by the model. The first, called the Markov assumption, states that the current state is dependent only on the previous state, this represents the memory of the model:

$$P(q_t | q_1^{t-1}) = P(q_t | q_{t-1}) \tag{2}$$

The independent assumption states that the output observation at time t is dependent only on the current state, it is independent of previous observations and states:

$$P(o_t | o_1^{t-1}, q_1^t) = P(o_t | q_t) \tag{3}$$

The problem of HMMs is to determine a method to adjust the model parameters (A, B, π) to maximize the probability of the observation sequence given the model. There is no known way to analytically solve for the model which maximize the probability of the observation sequence. In fact, given finite observation sequence as training data, there is no optimal way of estimating the model parameters[9]. We can, however, choose $\lambda = (A, B, \pi)$ such as $P(O|\lambda)$ is locally maximized using an iterative procedure such as Baum-Welch method(or equivalently the EM method) [10], or using gradient techniques.

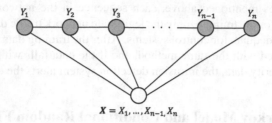

Fig. 1. Graphical structure of a chain-structured CRFs for sequences. The variables corresponding to unshaded nodes are not generated by the model.

3.2   CRF

A CRF may be viewed as an undirected graphical model, or Markov random field, globally conditioned on $\mathbf{X}$, the random variable representing observation sequences. Formally, we define $G = (V, E)$ to be an undirected graph such that there is a node $v \in V$ corresponding to each of the random variables representing an element $Y_v\, of\, \mathbf{Y}$. If each random variable Y_v obeys the Markov property with respect to G, then $(\mathbf{Y}, \mathbf{X})$ is a conditional random field. In theory the structure of graph G may be arbitrary, provided it represents the conditional independence in the label sequences being modeled. However, when modeling sequences, the simplest and most common graph structure encountered is that in which the nodes corresponding to elements of Y form a simple first-order chain, as illustrated in Figure 1.

CRF is a probabilistic model which has a graphical structure. The input data is $\mathbf{x}$ and the estimated output data is $\mathbf{y}$. The general graphical models estimate a joint probabilistic distribution $P(\mathbf{x}, \mathbf{y})$. The general graphical models need to estimate a probabilistic distribution $P(\mathbf{x})$ to estimate the joint probabilistic distribution $P(\mathbf{x}, \mathbf{y})$. If the relation between input data is complicated, it is sometimes hard to estimate the joint probabilistic distribution. CRF estimates a conditional distribution $P(\mathbf{y}|\mathbf{x})$ directly.

CRF is computed by the following. The input data sequence set is X and the output label sequence set is Y. The input data sequence $x = \{x_1, \cdots, x_K\} \in X$ and and the output label sequence $\mathbf{y} = \{y_1, \cdots, y_K\} \in Y$ consists of K elements. The conditional probabilistic distribution $P(\mathbf{y}|\mathbf{x})$ is defined by the next equation.

$$P(\mathbf{y}|\mathbf{x}) = \frac{1}{Z(\mathbf{x})} \exp \left\{ \sum_{k=1}^{K} \lambda_k f_k(\mathbf{x}_t, \mathbf{y}_t) \right\} \qquad (4)$$

where $\Lambda = \{\lambda_k, k = 1, \cdots, K\}$ is a set of parameters. $f_k(\mathbf{x}_t, \mathbf{y}_t)$ denotes a characteristic function and is represented by the next equation.

$$\text{If } \mathbf{y}_t = \mathbf{y}_k, \quad f_k(\mathbf{x}_t, \mathbf{y}_t) = \mathbf{x}_k \quad \text{and if } \mathbf{y}_t \neq \mathbf{y}_k, \quad f_k(\mathbf{x}_t, \mathbf{y}_t) = 0 \qquad (5)$$

$Z(\mathbf{x})$ is a partition function and represented by the following equation.

$$Z(\mathbf{x}) = \sum_{\mathbf{y}} \exp \left\{ \sum_{k=1}^{K} \lambda_k f_k(\mathbf{x}, \mathbf{y}) \right\} \qquad (6)$$

where $\mathbf{x}^{(i)} = \{x_1^{(i)}, \cdots, x_T^{(i)}\}, i = 1, \cdots, N$ is a sequence of the input data and to $\mathbf{y}^{(i)} = \{y_1^{(i)}, \cdots, y_T^{(i)}\}, i = 1, \cdots, N$ and N is the number of data to generate a probabilistic model. Finally, the parameter λ is estimated by the maximum likelihood estimation based on the following likelihood function[12].

$$l(\lambda_i) = \sum_{i=1}^{N} \log P(\mathbf{y}^{(i)}|\mathbf{x}^{(i)}) \tag{7}$$

where $\mathbf{y}^{(i)}, \mathbf{x}^{(i)}$ are a set of training data $D = (\mathbf{x}^{(i)}, \mathbf{y}^{(i)})_{i=1}^{N}$ to generate probabilistic models[11].

4 Experiments

4.1 Experimental Setup

Our experiments used two kinds of communication data. One is benchmark data which are supplied by DARPA[13,14]. Another one is communication data of a model control system. The data are generated by our model system. Now, this paper introduces these two kinds of communication data.

Benchmark Data. In order to evaluate intrusion detection methods for an information system, our experiments adopted bench DARPA benchmark data[13,14]. From DARPA benchmark data, we randomly pick out 10, 000 packets of normal communication data for training probabilistic models of normal communication sequences. And from DARPA benchmark data, we randomly pick out 10, 000 packets of communication data, which include abnormal communication data, for evaluating the performance of some intrusion detection methods. Our experiments generated ten training datasets and ten test datasets. The features used as input data are shown in the table 1.

Control System Communication Data. The model control system communication network used in the experiment is shown in Fig. 2 and was developed in our laboratory. Japanese electric power companies have two communication networks, which are the control system network and business network, which is defined as a type of business social network whose reason for existing is business networking activity. And these communication networks are disconnected physically. Therefore, it is assumed here that no connections exist between our control system network and the corporate business network. All the equipment, applications and security measures in the figure belong solely to the control system network, and the placement and settings mentioned are irrelevant to the corporate business network. The control system master server located in the control center is the direct control and data acquisition interface to the control system equipment, or field devices. To control the field devices located in the substation, the master server communicates with the server in the substation, which in turn sends the actual control signals to the field device. In this model system, the control system field devices are emulated in the same terminal as the substation server. The

Table 1. Features of Input Data for DARPA data

Features
-Source IP address (IP address, MAC address)
-Destination IP address (IP address, MAC address)
-Source port number
-Destination port number
-Protocol
-Data length
-Sequence number
-Identification number of field device
-Type of control command or state information
-Interval between packets
-Interval between use of field device
-Interval between control command or state information retrieval
-Frequency of use of field device

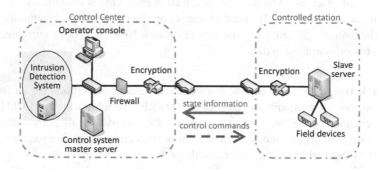

Fig. 2. An Overview of the Model System used for the experiment
Encryption is the process of encoding messages or information

state information of the field devices and any responses to the control commands are sent through the substation server to the control system master server, and then sent to any communicating operator consoles, which in this case are located within the control center.

Using the model control system, we collected 10,000 packets of normal communication data for training two probabilistic models. The features used from the data are shown in Table 2. The data was calibrated based on prior knowledge of the control system, to have values closer to 0 for less probable data.

Using the model control system, we collected 500 packets of unauthorized communication data for evaluating probabilistic models as intrusion detection systems. The unauthorized communication data are generated independently of the normal communication data by simulated cyber attacks, because we do not have any real cyber attacks for control systems. The model control system generated three types of the virtual cyber attacks. The three types of the virtual cyber attacks are the following.

Table 2. Data used for Training

Data	Description
-Source IP address(IP address, MAC address) -Destination IP address(IP address, MAC address)	-Addresses belonging to devices in the system: 1 -Addresses belonging to the same network as the system: 0.5 -Addresses belonging to different network: 0
-Source port number -Destination port number	-Port numbers used in the system: 1 -Port numbers reserved for use in the system: 0.5 -Port numbers not expected to be used in the system: 0
-Protocol -Identification number of field device -Type of control command or state information	-Content specified to be used in the system: 1 -Content not clearly specified to be used in the system: 0.5 -Content clearly specified not to be used in the system: 0
-Data length	-Length within the range used in the system: 1 -Length out of the range used in the system: 0.5
-Interval between packets	-Time elapsed after previous packet
-Interval between use of field device	-Time elapsed after use of a certain field device
-Interval between control command or state information retrieval	-Time elapsed after use of the same control command or state information retrieval
-Frequency of use of field device	-Frequency of usage of a certain field device within the training dataset

Type 1: Operational Commands for Nonexistent Controlled Equipments. This virtual cyber attack uses a controlled equipment ID which does not exit in the control system. In this case, an attacker does not have enough knowledge of the control system.

Type 2: Iterative Selection of Different Controlled Equipments to Virtually Operate. In this virtual cyber attack, a controlled equipment ID is a regularized ID. But the virtual cyber attack selects different controlled equipments iteratively. In this case, an attacker does not have enough knowledge of the control system and explore the control system architecture.

Type 3: Iterative Operational Commands for a Controlled Equipment. In this virtual cyber attack, operational commands are regularized commands. But this cyber attack sends regularized operational commands to a controlled equipment iteratively. In this case, the attacker wants to give a damage to controlled equipments.

4.2   Evaluation Criteria

In our experiments, we use the precision and the recall to evaluate the performance of the intrusion detection. The precision is represented by the next equation.

$$\text{Precision} = \frac{\text{True Positive}}{(\text{True Positive}) + (\text{False Positive})} \tag{8}$$

The recall is represented by the next equation.

$$\text{Recall} = \frac{\text{True Positive}}{(\text{True Positive}) + (\text{False Negative})} \tag{9}$$

And we also use the F-measure to evaluate the performance and the F-measure is represented by the next equation.

$$\text{F-measure} = \frac{(1 + \beta^2) \times (\text{Precision}) \times (\text{Recall})}{\beta^2 \times (\text{Precision} + \text{Recall})} \tag{10}$$

where, β denotes a relative importance between the precision and the recall and is usually set up 1[15]. Therefore, we also set β up 1 in our experiments.

Table 3. Precision, Recall and F-measure for different methods for DARPA data: Length of sequence denotes CRF probabilistic model. And 2, 3 and 4 denotes the length of the considered sequence.

Method	Precision	Recall	F-measure
Without sequences	0.8820	0.9465	0.9131
Rule based sequences	0.9034	0.8887	0.8960
HMM	0.8902	0.9180	0.9039
Length of sequence: 2	0.9021	0.9142	0.9207
Length of sequence: 3	0.9195	0.9219	*0.9216*
Length of sequence: 4	0.9189	0.9178	0.9183

4.3 Experimental Results

This section presents two kinds of experimental results. One experimental results are based on DARPA benchmark data. Another experimental results are based on control system communication data.

Benchmark Data(DARPA Data). In our experiments, we apply training data, which are pick out $10,000$ packets of normal communication data from DARPA data, and test data, which are pick out $10,000$ packets of communication data, which include unauthorized communication data from DARPA data, and two probabilistic models such as HMM and CRF. The table 3 shows the precision, recall and F-measure of HMM and CRF and CRF without sequences information, and an empirical expertise of control systems(Rule based sequences)[16].

The intrusion detection method without sequences shows that the performance of recall is high but the performance of precision is not high. The CRF based intrusion detection method shows the best performance of precision, recall and F-measure when the number of input sequences is four. The performance of the CRF based intrusion detection method is as same as the other methods for DARPA data.

Control System Communication Data. In our experiments, we apply training data, which are collected $10,000$ packets of normal communication data from the model control system, and test data, which are collected 500 packets of unauthorized communication data from the model control system, two probabilistic models such as HMM and CRF. The table 4 shows the precision, recall and F-measure of HMM and CRF. and CRF without sequences information, and an empirical expertise of control systems(Rule based sequences) [16].

From the table 4, CRF without sequences could not detect typical intrusion of control systems. The methods based the control sequences showed the higher performance than the CRF without sequences. The two probabilistic modes achieved the higher performance than the empirical expertise. In our experiments, CRF showed the higher performance than the other methods. Especially, when the length of the considered sequence is 3, CRF achieved the highest performance.

The proposed intrusion detection method based on CRF could detect most of intrusions in the model control system. But our proposed method could not detect few

Table 4. Precision, Recall and F-measure for different methods for control system communication data: Length of sequence denotes CRF probabilistic model. And 2, 3 and 4 denotes the length of the considered sequence.

Method	Precision	Recall	F-measure
Without sequences	0.9016	0.8869	0.8942
Rule based sequences	0.9451	0.8931	0.9184
HMM	0.9523	0.9089	0.9301
Length of sequence: 2	0.9713	0.9198	0.9448
Length of sequence: 3	0.9839	0.9266	*0.9544*
Length of sequence: 4	0.9727	0.9279	0.9498

intrusions. In these few intrusions, an attacker included into the control system and made regular controls. It is impossible to detect intrusions by monitoring only control communication sequences when the attacker makes regular controls.

In our experiments, the performance of intrusion detection depends on the length of sequences. The network intrusion detection system has high frequent control sequences, which are generated by the monitored control system. So, the performance of intrusion detection depends on the length of sequences.

5   Conclusion

In this paper, we compared algorithms that reflect the sequence characteristics of network traffic into the intrusion detection system. The learning only utilizes normal data, assuming that there is no prior knowledge on attacks in the system. The results show that the approach successfully identifies the sequences inherent in the system for the intrusion detection. With this method, it is possible for the intrusion detection system to detect attacks that deviate from the typical system actions. In addition, this method does not need prior knowledge of attacks, which is the case for new, unknown cyber attacks.

Especially, the CRF based method could achieved better performance than any other methods in our experiments. But the method could not detect few intrusions. In these few intrusions, an attacker included into the control system and made regular controls. It is impossible to detect intrusions by monitoring only control communication sequences when the attacker makes regular controls. We need to use another additional information to detect these intrusion.

Finally, the paper showed two important things.

1. The experimental results of our proposed methods show that sequence information in control systems is very important for detecting some intrusion attacks.
2. We can expect that an intrusion detection method based on sequence information in control systems will become a real world application in the near future.

This method may have difficulties in computation size with a large system, as it becomes necessary to retain a large number of features, resulting in large memory requirements. Future work includes the consideration of reducing computation by machine learning techniques.

References

1. National Security Agency: Defense in Depth: A practical strategy for achieving Information Assurance in today's highly networked environments,
   `http://www.nsa.gov/ia/-files/support/defenseindepth.pdf`
2. SANS Institute: Intrusion Detection FAQ,
   `http://www.sans.org/resources/idfaq`
3. Cheung, S., Dutertre, B., Fong, M., Lindqvist, U., Skinner, K., Valdes, A.: Using Model-based Intrusion Detection for SCADA Networks. In: Proc. of the SCADA Security Scientific Symposium (January 2007)
4. Moran, B., Belisle, R.: Modeling Flow Information and Other Control System Behavior to Detect Anomalies. In: Proc. of the SCADA Security Scientific Symposium (January 2008)
5. Kiuchi, M., Serizawa, Y.: Security Technologies, Usage and Guidelines in SCADA System Networks. In: ICCAS-SICE (2009)
6. Onoda, T., Kiuchi, M.: Analysis of Intrusion Detection in Control System Communication Based on Outlier Detection with One-Class Classifiers. In: Huang, T., Zeng, Z., Li, C., Leung, C.S. (eds.) ICONIP 2012, Part V. LNCS, vol. 7667, pp. 275–282. Springer, Heidelberg (2012)
7. Osareh, A., Shadgar, B.: Intrusion Detection in Computer Networks based on Machine Learning Algorithms. International Journal of Computer Science and Network Security 8(11) (November 2008)
8. Chandola, V., Banerjee, A., Kumar, V.: Outlier Detection: A Survey, University of Minnesota Technical Report TR 07-017
9. Rabiner, L.R.: A tutorial on hidden Markov models and selected applications in speech recognition. Proceedings of the IEEE 77(2), 257–285 (1989)
10. Baum, L.E., Petrie, T., Soules, G., Weiss, N.: A maximization technique occurring in the statistical analysis of probabilistic functions of markov chains. The Annals of Mathematical Statistics 41(1), 164–171 (1970)
11. Lafferty, J., McCallum, A., Pereira, F.: Conditional random fields: probabilistic models for segmenting and labeling sequence data. In: International Conference on Machine Learning (2001)
12. CRF++: Yet Another CRF toolkit, `http://crfpp.sourceforge.net/`
13. Lippmann, R.P., Haines, J.W., Fried, D.J., Korba, J., Das, K.: The 1999 DARPA off-line intrusion detection evaluation. Computer Networks 34, 579–595 (2000)
14. DARPA: Intrusion Detection evaluation data-set, `http://www.ll.mit.edu/mission/communications/ist/CST/index.html`
15. Zhang, D., Leckie, C.: An Evaluation Technique for Network Intrusion Detection Systems. In: Proc. of the 1st International Conference on Scalable Information Systems, InfoScale 2006 (2006)
16. Kiuchi, M., Ohba, E., Serizawa, Y.: Customizing Control System Intrusion Detection at the Application Layer. In: Proc. of the SCADA Security Scientific Symposium 2009, Digital Bond Press (January 2009)

Compressive ELM: Improved Models through Exploiting Time-Accuracy Trade-Offs

Mark van Heeswijk[1], Amaury Lendasse[1,2,3], and Yoan Miche[1]

[1] Aalto University School of Science,
Department of Information and Computer Science,
P.O. Box 15400, FI-00076 Aalto, Finland
[2] Arcada University of Applied Sciences, Helsinki, Finland
[3] Department of Mechanical and Industrial Engineering, The University of Iowa,
Iowa City, IA 52242-1527, USA

Abstract. In the training of neural networks, there often exists a trade-off between the time spent optimizing the model under investigation, and its final performance. Ideally, an optimization algorithm finds the model that has best test accuracy from the hypothesis space as fast as possible, and this model is efficient to evaluate at test time as well. However, in practice, there exists a trade-off between training time, testing time and testing accuracy, and the optimal trade-off depends on the user's requirements. This paper proposes the Compressive Extreme Learning Machine, which allows for a time-accuracy trade-off by training the model in a reduced space. Experiments indicate that this trade-off is efficient in the sense that on average more time can be saved than accuracy lost. Therefore, it provides a mechanism that can yield better models in less time.

Keywords: Extreme Learning Machine, ELM, random projection, compressive sensing, Johnson-Lindenstrauss, approximate matrix decompositions.

1 Introduction

When choosing a model for solving a machine learning problem, which model is most suitable depends a lot on the context and the requirements of the application. For example, it might be the case that the model is trained on a continuous stream of data, and therefore has some restrictions on the training time. On the other hand, computational time in the testing phase might be restricted, like in a setting where the model is used as the controller for an aircraft or a similar setting that requires fast predictions. Alternatively, the context in which the model is applied might not have any strong constraints on the computational time, and above all, accuracy or interpretability is considered most important regardless of the computational time.

This paper focuses on time-accuracy trade-offs in a neural network architecture known as Extreme Learning Machine [1], and on trade-offs between training time and accuracy in particular. This trade-off can be affected in two ways:

V. Mladenov et al. (Eds.): EANN 2014, CCIS 459, pp. 165–174, 2014.
© Springer International Publishing Switzerland 2014

- by improving the accuracy through spending more time optimizing the model,
- or vice-versa, by reducing the computational time of the model, without sacrificing accuracy too much.

Each type of model has its own ways of balancing computational time and accuracy, and has an associated curve (or set of points) on a "training time"-accuracy plot that expresses the efficiency of the model in achieving a certain accuracy (the closer the curve is to the bottom left, the better). Thus, given a collection of models, the question becomes: which model produces the best accuracy the fastest?

The remainder of this paper is organized as follows. Section 2 discusses the preliminaries and methods relevant for this paper and gives an example of the time-accuracy trade-offs that exist within several ELM variants. This illustrates the notion of 'efficiency' of a model, and motivates the choice of model that is studied in the rest of the paper. Section 3 proposes the Compressive ELM, a new model which allows trading off computational time and accuracy by performing the training in a reduced problem space rather than the original space. Finally, Section 4 contains the experiments and analysis which form the validation for the proposed approach.

2   Background

Regression / Classification. In this paper, the focus is on the problem of regression, which is about establishing a relationship between a set of output variables (continuous) $y_i \in \mathbb{R}, 1 \leq i \leq M$ (single-output here) and another set of input variables $\mathbf{x}_i = (x_i^1, \ldots, x_i^d) \in \mathbb{R}^d$. Note that although in this paper the focus is on regression, the proposed approach can just as well be used when applying the ELM in a classification context.

Extreme Learning Machine (ELM). The ELM algorithm is proposed by Huang *et al.* in [1] and uses Single-Layer Feedforward Neural Networks (SLFN). The key idea of ELM is that the hidden layer weights and hidden layer biases of the SLFN can be generated randomly, and do not need to be trained.

Consider a set of N distinct samples $(\mathbf{x}_i, y_i)$ with $\mathbf{x}_i \in \mathbb{R}^d$ and $y_i \in \mathbb{R}$. Then, an SLFN with M hidden neurons can be written as

$$\sum_{i=1}^{M} \beta_i f(\mathbf{w}_i \cdot \mathbf{x}_j + b_i), j \in [1, N], \tag{1}$$

with f being the transfer function, $\mathbf{w}_i$ the input weights to the ith neuron in the hidden layer, b_i the hidden layer biases and β_i the output weights.

Gathering the outputs of the transfer functions in an $N \times M$ matrix $\mathbf{H}$ and the targets in $\mathbf{Y}$, in case the network would perfectly approximate the targets this can be written compactly as

$$\mathbf{H}\beta = \mathbf{Y}, \tag{2}$$

where $\mathbf{H}$ is the hidden layer output matrix defined as

$$\mathbf{H} = \begin{pmatrix} f(\mathbf{w}_1 \cdot \mathbf{x}_1 + b_1) & \cdots & f(\mathbf{w}_M \cdot \mathbf{x}_1 + b_M) \\ \vdots & \ddots & \vdots \\ f(\mathbf{w}_1 \cdot \mathbf{x}_N + b_1) & \cdots & f(\mathbf{w}_M \cdot \mathbf{x}_N + b_M) \end{pmatrix} \qquad (3)$$

and $\beta = (\beta_1 \ldots \beta_M)^T$ and $\mathbf{Y} = (y_1 \ldots y_N)^T$. Under the condition that the input weights and biases are randomly initialized, and the transfer function f is a bounded non-constant piecewise continuous activation function, [2] proves that the ELM is a universal approximator. Therefore, given enough neurons, the ELM can approximate a function or set of target values as good as desired. The optimal least-squares solution to the equation $\mathbf{H}\beta = \mathbf{Y}$ in the ELM algorithm is $\beta = \mathbf{H}^\dagger \mathbf{Y}$, where $\mathbf{H}^\dagger$ is the pseudo-inverse of $\mathbf{H}$. In summary then, the standard ELM algorithm can be described in Algorithm 1. Theoretical proofs and a more thorough presentation of the ELM algorithm can be found in [1].

Algorithm 1. Standard ELM

Given a training set $(\mathbf{x}_i, y_i), \mathbf{x}_i \in \mathbb{R}^d, y_i \in \mathbb{R}$, an activation function $f : \mathbb{R} \mapsto \mathbb{R}$ and M hidden nodes:

1: - Randomly assign input weights $\mathbf{w}_i$ and biases $b_i, i \in [1, M]$;
2: - Calculate the hidden layer output matrix $\mathbf{H}$;
3: - Calculate output weights matrix $\beta = (\mathbf{H}^T\mathbf{H})^{-1}\mathbf{H}\mathbf{Y} = \mathbf{H}^\dagger\mathbf{Y}$.

Efficient Optimization of Regularization Parameter with SVD. Trained on a limited number of samples, the standard ELM is prone to overfitting the training data. One way of preventing overfitting is by applying Tikhonov Regularization, in which case pseudo-inverse used in the ELM becomes

$$\mathbf{H}^\dagger = (\mathbf{H}^T\mathbf{H} + \lambda\mathbf{I})^{-1}\mathbf{H}^T$$

for some regularization parameter λ [3]. Each value of λ results in a different pseudo-inverse $\mathbf{H}^\dagger$, and it would be computationally expensive to recompute the pseudo-inverse for every λ. However, by incorporating the regularization in the singular value decomposition (SVD) approach to compute the pseudo-inverse, it becomes possible to obtain the various $\mathbf{H}^\dagger$'s with minimal re-computation [4]. This scheme is first described in the context of ELM in [5], and is summarized next (with some minor optimizations). Suppose

$$\begin{aligned}
\hat{\mathbf{Y}} &= \mathbf{H}\beta \\
&= \mathbf{H}(\mathbf{H}^T\mathbf{H} + \lambda\mathbf{I})^{-1}\mathbf{H}^T\mathbf{Y} \\
&= \mathbf{H}\mathbf{V}(\mathbf{D}^2 + \lambda\mathbf{I})^{-1}\mathbf{D}\mathbf{U}^T\mathbf{Y} \\
&= \mathbf{U}\mathbf{D}\mathbf{V}^T\mathbf{V}(\mathbf{D}^2 + \lambda\mathbf{I})^{-1}\mathbf{D}\mathbf{U}^T\mathbf{Y} \\
&= \mathbf{U}\mathbf{D}(\mathbf{D}^2 + \lambda\mathbf{I})^{-1}\mathbf{D}\mathbf{U}^T\mathbf{Y} \\
&= \mathrm{HAT} \cdot \mathbf{Y}
\end{aligned}$$

where $\mathbf{D}(\mathbf{D}^2 + \lambda\mathbf{I})^{-1}\mathbf{D}$ is a diagonal matrix with $\frac{d_{ii}^2}{d_{ii}^2+\lambda}$ as the ith diagonal entry. From the above equations it can now be seen that given $\mathbf{U}$:

$$\mathrm{MSE}^{\mathrm{TR\text{-}PRESS}} = \frac{1}{N}\sum_{i=1}^{N}\left(\frac{y_i - \hat{y}_i}{1 - hat_{ii}}\right)^2$$

$$= \frac{1}{N}\sum_{i=1}^{N}\left(\frac{y_i - \hat{y}_i}{1 - \mathbf{h}_{i\cdot}(\mathbf{H}^T\mathbf{H} + \lambda\mathbf{I})^{-1}\mathbf{h}_{i\cdot}^T}\right)^2$$

$$= \frac{1}{N}\sum_{i=1}^{N}\left(\frac{y_i - \hat{y}_i}{1 - \mathbf{u}_{i\cdot}\left(\frac{d_{ii}^2}{d_{ii}^2+\lambda}\right)\mathbf{u}_{i\cdot}^T}\right)^2$$

where $\mathbf{h}_{i\cdot}$ and $\mathbf{u}_{i\cdot}$ are the ith row vectors of $\mathbf{H}$ and $\mathbf{U}$, respectively. The optimal Tikhonov-regularized PRESS and corresponding λ can be determined efficiently using Algorithm 2. Due to the convex nature of criterion $\mathrm{MSE}^{\mathrm{TR\text{-}PRESS}}$ with respect to regularization parameter λ, the Nelder-Mead procedure used for optimizing λ converges quickly in practice [6,7].

Algorithm 2. Tikhonov-regularized PRESS. In practice, the **while** part of this algorithm (convergence for λ) is solved using by a Nelder-Mead approach [6], a.k.a. downhill simplex.

1: Decompose $\mathbf{H}$ by SVD: $\mathbf{H} = \mathbf{UDV}^T$
2: Precompute $\mathbf{B} = \mathbf{U}^T\mathbf{y}$
3: **while** no convergence on λ achieved **do**
4:    - Precompute $\mathbf{C} = \mathbf{U}\cdot\mathrm{diag}\left(\frac{d_{11}^2}{d_{11}^2+\lambda}, \ldots, \frac{d_{nn}^2}{d_{nn}^2+\lambda}\right)$
5:    - Compute $\hat{\mathbf{y}} = \mathbf{CB}$, the vector containing all $\hat{y}_i$
6:    - Compute $\mathbf{d} = \mathrm{diag}\left(\mathbf{CU}^T\right)$, the diagonal of the HAT matrix, by taking the row-wise dot-product of $\mathbf{C}$ and $\mathbf{U}$
7:    - Compute $\varepsilon = \frac{\mathbf{y}-\hat{\mathbf{y}}}{1-\mathbf{d}}$, the leave-one-out errors
8:    - Compute $\mathrm{MSE}^{\mathrm{TR\text{-}PRESS}} = \frac{1}{N}\sum_{i=1}^{N}\varepsilon_i^2$
9: **end while**
10: Keep the best $\mathrm{MSE}^{\mathrm{TR\text{-}PRESS}}$ and the associated λ value

---

Example: Time-Accuracy Trade-offs for Several ELM Variants. In order to illustrate what time-accuracy trade-offs exist within ELM, and to motivate the choice of model studied later in this paper, this section presents time-accuracy trade-offs of several models:

- **ELM:** the basic ELM [1].
- **Optimally Pruned ELM (OP-ELM):** ELM trained by generating a set of neurons, ranking them by relevance, and then determining the optimal prefix of that sorted list of neurons in terms of leave-one-out error [8]
- **TROP-ELM:** OP-ELM with efficient optimization of the Tikhonov regularization integrated, using the SVD approach to computing $\mathbf{H}^\dagger$ [5]

- **TR-ELM:** Tikhonov-regularized ELM [3], with efficient optimization of regularization parameter λ, using the SVD approach. [9]
- **BIP(0.2), BIP(rand), BIP(CV):** ELMs pretrained using Batch Intrinsic Plasticity mechanism [10], aimed at adapting the hidden layer weights and biases, such that they retain as much information from the input as possible. The variants included here have the BIP parameter μ_{exp} fixed to a 0.2, randomized, or cross-validated over 20 possible values.

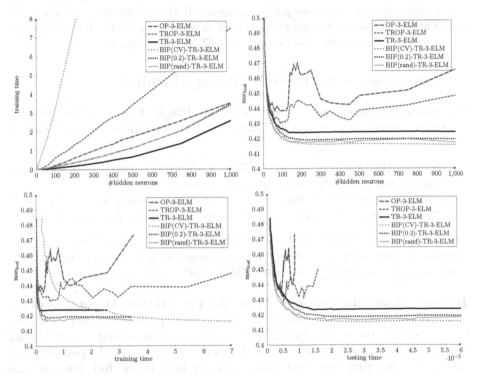

Fig. 1. Results for various ELM variants on Abalone UCI data set

All these models are trained and tested on the Abalone data set from the UCI repository [11] (see Section 4 for details), use ternary weights (see [9]), and have an initial number of hidden neurons varying between 2 and 1000. Each method trains and optimizes the ELM in its own way, with results as summarized in Figure 1. Depending on the users criteria, these results suggest:

- if *training time* most important, then BIP(rand)-TR-3-ELM is the obvious choice from all candidates as it provides almost optimal performance, while keeping training time low.
- if *test error* is most important, then BIP(CV)-TR-3-ELM is the best choice. However, since it cross-validates over 20 possible parameter values, the training time is 20 times as high, while only giving slightly better accuracy.

– if *testing time* is most important, then surprisingly TR-3-ELM is also the most attractive model. Even though OP-ELM and TROP-ELM tend to be faster in test, they suffer from slight overfitting as the number of initial hidden neurons increases. Therefore, the TR-3-ELM is the best choice, since it generally results in models with the best accuracy and lowest testing time.

Since TR-ELM offers attractive trade-offs between speed and accuracy, this model will be central in the rest of the paper. Furthermore, since due to the proper regularization the TR-ELM does not seem to overfit even for large number of neurons: more neurons generally means better accuracy. Naturally, this comes at an increase in training time, which is something that will be addressed in the next section, where the Compressive ELM is presented.

3   Compressive Extreme Learning Machine

Considering training time-accuracy trade-offs like in Figure 1, two possible strategies present itself to obtain models that are preferable over other models:

– reducing test error, using some efficient algorithm ("in terms of training time-accuracy plot: "pushing the curve down")
– reducing computational time, while retaining as much accuracy as possible ("in terms of training time-accuracy plot: "pushing the curve to the left")

The latter is the strategy that is taken in Compressive ELM: instead of performing the training in the original problem space, it performs the training in a reduced space, and then project the solution back to the original space.

Johnson-Lindenstrauss and Approximate Matrix Decompositions. Given an $m \times n$ matrix, an approximate matrix decomposition can be achieved by first embedding the rows of the matrix into a lower-dimensional space (through one of many available low-distortion Johnson-Lindenstrauss-like embeddings), solving the decomposition, and then projecting back to the full space. If such an embedding (or sketch) is accurate, then this allows for solving the problem with high accuracy in reduced time. The algorithm for Approximate SVD is summarized in Algorithm 3, and more background can be found in [12].

Algorithm 3. Approximate SVD [12]

Given an $m \times n$ matrix A, compute k-term approximate SVD $A \approx UDV^T$ as follows:

1: - Form the $n \times (k+p)$ random matrix Ω. (where p is small over sampling parameter)

2: - Form the m $\times (k + p)$ sampling matrix $Y = A\Omega$. ("sketch" it by applying Ω)
3: - Form the m $\times (k + p)$ orthonormal matrix Q, such that $range(Q) = range(Y)$.
4: - Compute $B = Q^*A$.
5: - Form the SVD of B so that $B = \hat{U}DV^T$
6: - Compute the matrix $U = Q\hat{U}$

Faster Sketching. Typically, the bottleneck in Algorithm 3 is the time it takes to sketch the matrix. Rather than using a class of random matrices of Gaussian variables for sketching A, one can also use random matrices that are sparse or structured in some way [13,14], for which the matrix-vector product can be computed more efficiently. Furthermore, Ailon and Chazelle [15] introduced the Fast Johnson-Lindenstrauss Transform (FJLT), which uses a class of random matrices that allow application of an $n \times n$ matrix to a vector in $\mathcal{O}(n \log(n))$, rather than the usual $\mathcal{O}(n^2)$. Besides this obvious speedup, this class of matrices is also more successful in creating a low-distortion embedding when applied to a sparse matrix. These transforms consist of the application of three easy-to-compute matrices

$$(P)_{k \times n} (H)_{n \times n} (D)_{n \times n}$$

where P, H, and D vary depending on the exact scheme. Generally, D is a diagonal matrix with random Rademacher variables $(-1, +1)$ on the diagonal, H is encoding either the discrete Hadamard or discrete Fourier transform, and P is a sparse random matrix or a matrix sampling random columns from H. The D matrix can be applied to a vector x in $\mathcal{O}(n)$, The H matrix can be applied in $\mathcal{O}(n \log(n))$, and the P matrix adds a factor $nnz(P)$ or k, depending on the type.

4   Experiments

This section describes the experiments that investigate the trade-off between computational time (both training and test), and the accuracy of the Compressive ELM in relation to, the dimensionality of the space into which the problem is reduced, using the sketch. For sketching, TR-3-ELMs with the following sketching schemes are considered, and compared with the standard TR-3-ELM:

- Gaussian: sketching is performed using a $k \times n$ matrix of random Gaussian variables
- FJLT: the transform introduced in [15], for which P is a sparse matrix of random Gaussian variables, and H encodes the Discrete Hadamard Transform
- SRHT: a variant of the FJLT, for which P is a matrix selecting k random columns from H, and H encodes the Discrete Hadamard Transform

The number of hidden neurons in each model is varied between 2 and 1000, and parameter k is chosen from $[50, 100, 200, 400, 600]$. Experiments are repeated with 200 random realizations of the training and test set, and average results over those 200 runs are reported.

Data and Preprocessing. As data sets, different regression tasks from the UCI machine learning repository [11] are tested. Due to space restrictions only the results for CaliforniaHousing and FJLT sketching are presented here, but similar results hold for the other data sets and sketching methods. In each run, the data is divided randomly into 8000 random samples for training and and the remaining 12640 samples for testing. The data is preprocessed in such a way that each input and output variable is zero mean and unit variance.

Results. The results of the experiment are summarized in Figure 2. There, it can be seen that

- setting k lower than the number of neurons results in faster training times (which makes sense since the problem solved is smaller).
- as long as parameter k is chosen large enough, the method is not losing efficiency (i.e. there is no model that achieves better error in the same computational time), and it is potentially gaining efficiency (as shown by the bottom-left plot of Figure 2.

Finally, the experiments showed that sketches with Gaussian matrices are generally the fastest. Furthermore, for the tested problem sizes, the SRHT (which allows an efficient matrix multiplication) is generally faster than the FJLT (which uses sparse matrices). Although for this problem size the SRHT and FJLT are slower, they might still be needed in case the matrix to sketch is sparse [15].

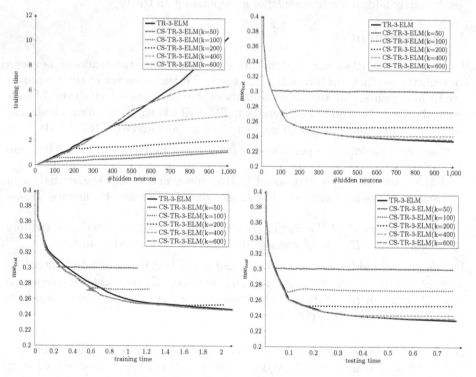

Fig. 2. Results for Compressive ELMs using FJLT sketching with varying k on CaliforniaHousing UCI data set

5   Conclusion

In this paper, the trade-off between computational time and test error has been investigated, in particular the trade-off between training time and test error. Having information about this trade-off for different models is useful information in selecting the most suitable model for a particular task.

The Compressive ELM proposed in this paper investigates a way to reduce training time by doing the optimization in a reduced space of k dimensions, and is shown to be efficient in the sense that (given k large enough), among the tested models the Compressive ELM achieves the best test error for each computational time (i.e. there are no models that achieve better test error and can be trained in the same or less time). A promising candidate for setting k such that it optimally reduces computational time (yet retains accuracy), would be to let k be informed by the theoretical bounds currently known for the sketching schemes. These theoretical bounds give lower bounds on k for which a low-distortion embedding of the given n points can be achieved with high probability. Although these bounds are typically not sharp (and therefore not optimal), in case the minimal k for successful embedding is lower than the number of neurons in the ELM, it can be exploited to reduce the training time.

Finally, developing low-distortion embeddings and sharpening their associated bounds is currently a hot topic of research, and any new developments in this area can easily be integrated to improve the performance of Compressive ELM.

References

1. Huang, G.-B., Zhu, Q.-Y., Siew, C.-K.: Extreme learning machine: Theory and applications. Neurocomputing 70(1-3), 489–501 (2006)
2. Huang, G.-B., Chen, L., Siew, C.-K.: Universal Approximation Using Incremental Constructive Feedforward Networks with Random Hidden Nodes. IEEE Transactions on Neural Networks 17(4), 879–892 (2006)
3. Deng, W.-Y., Zheng, Q.-H., Chen, L.: Regularized extreme learning machine. In: IEEE Symposium on Computational Intelligence and Data Mining, CIDM 2009, pp. 389–395 (2009)
4. van Heeswijk, M., Miche, Y., Oja, E., Lendasse, A.: GPU-accelerated and parallelized ELM ensembles for large-scale regression. Neurocomputing 74(16), 2430–2437 (2011)
5. Miche, Y., van Heeswijk, M., Bas, P., Simula, O., Lendasse, A.: TROP-ELM: A double-regularized ELM using LARS and Tikhonov regularization. Neurocomputing 74(16), 2413–2421 (2011)
6. Nelder, J., Mead, R.: A simplex method for function minimization. The Computer Journal 7(4), 308–313 (1965)
7. Lagarias, J.C., Reeds, J.A., Wright, M.H., Wright, P.E.: Convergence Properties of the Nelder–Mead Simplex Method in Low Dimensions. SIAM Journal on Optimization 9, 112–147 (1998)
8. Miche, Y., Sorjamaa, A., Bas, P., Simula, O., Jutten, C., Lendasse, A.: OP-ELM: optimally pruned extreme learning machine. IEEE Transactions on Neural Networks 21(1), 158–162 (2010)

9. van Heeswijk, M., Miche, Y.: Binary/Ternary Extreme Learning Machines. Neurocomputing (to appear)
10. Neumann, K., Steil, J.J.: Batch intrinsic plasticity for extreme learning machines. In: Honkela, T. (ed.) ICANN 2011, Part I. LNCS, vol. 6791, pp. 339–346. Springer, Heidelberg (2011)
11. Asuncion, A., Newman, D.J.: UCI Machine Learning Repository (2007)
12. Halko, N., Martinsson, P.-G., Tropp, J.: Finding structure with randomness: Probabilistic algorithms for constructing approximate matrix decompositions (September 2011) arXiv:0909.4061
13. Achlioptas, D.: Database-friendly random projections: Johnson-Lindenstrauss with binary coins. Journal of Computer and System Sciences 66(4), 671–687 (2003)
14. Matoušek, J.: On variants of the Johnson-Lindenstrauss lemma. Random Structures & Algorithms, 142–156 (2008)
15. Ailon, N., Chazelle, B.: Approximate nearest neighbors and the fast Johnson-Lindenstrauss transform. In: Proceedings of the Thirty-Eighth Annual ACM Symposium on Theory of Computing, STOC 2006, pp. 557–563. ACM Press, New York (2006)

Detecting Port Scans against Mobile Devices with Neural Networks and Decision Trees

Christo Panchev, Petar Dobrev, and James Nicholson

Department of Computing, Engineering and Technology,
University of Sunderland,
Sunderland SR6 0RD, United Kingdom
christo.panchev@sunderland.ac.uk

Abstract. Recently, mobile devices such as smartphones and tablets have emerged as one of the most popular forms of communication. This trend raises the question about the security of the private data and communication of the people using those devices. With increased computational resources and versatility the number of security threats on such devices is growing rapidly. Therefore, it is vital for security specialists to find adequate anti-measures against the threats. Machine Learning approaches with their ability to learn from and adapt to their environments provide a promising approach to modelling and protecting against security threats on mobile devices. This paper presents a comparative study and implementation of Decision Trees and Neural Network models for the detection of port scanning showing the differences between the responses on a desktop platform and a mobile device and the ability of the Neural Network model to adapt to the different environment and computational resource available on a mobile platform.

Keywords: Intrusion Detection, Port Scanning, Cascade Correlation Neural Networks, Decision Trees, Android, Mobile devices.

1 Introduction

Recent studies have shown that the number of security threats is growing rapidly [1] - with many of the network enabled vulnerability exploitation attacked previously reserved for mainframe and desktop computers now being directed toward mobile devices. Often users tend to connect to different public networks in order to use their applications and their devices are exposed to the threat of port scanning. The intention behind this form of attack is to find ports which are open from an application running on the machine, identify and to exploit its vulnerabilities. This type of attack is well known by security specialists, and is usually captured by Intrusion Detection/Prevention System (IDS/IPS). Such systems use different types of techniques in order to be able to detect the attack and to alert about its existence. While existing IDS are relatively successful in detecting potential attacks against desktop and server environments, they are not directly portable onto mobile platforms. The main reason draws from the fact

V. Mladenov et al. (Eds.): EANN 2014, CCIS 459, pp. 175–182, 2014.

that different mobile platforms differ in the available computational resource, means and speed of connectivity.

Many types of Intrusion Detection Systems exist these days for servers and personal computers [2], however literature provides very few examples of such systems for mobile devices [3, 4] , therefore the aim of this project is to produce a prototype of an Intrusion Detection System, which will be able to capture and alert about port scans performed against mobile devices.

Intrusion Detection System is a system which gathers and analyses information from computer system or a network in order to find incidents such as security policies violations and attack attempts [3]. Different IDS employ different technologies or techniques in the process of gathering and analysing the information depending on specific factors, such as type and capabilities of the protected device.

IDS are broadly divided into two types - Network Based Intrusion Detection System (NIDS) and Host Based Intrusion Detection System (HIDS). In general NIDS is a system which listens to the traffic and analyses the packets which are crossing a computer network, while HIDS gathers information about internal processes in the system such as processor utilisation, memory utilization, system files integrity, etc. Therefore, the part of HIDS which gathers the data will be checking information from system calls or system log files [6]. The study presented here focuses on analysis the network traffic of a single host and does not make use of the host-available resource utilisation data, hence it can be categorised as a single host NIDS.

Further, IDS detection algorithms can be categorised into Anomaly Detection, Misuse Detection and Hybrid [7, 8]. Following the first IDS model by Dorothy Denning and Peter Neumann in 1987 [9], Anomaly Detection based IDS have been based on the assumption that any form of system abuse, should generate anomalies, therefore is detectable. Anomaly Detection algorithms operate into two phases: the first one involves the profiling of or training the normal behaviour by observation of the standard non-malicious traffic, the second one deals with the detection of the anomalies by comparing the profile created in the first phase to the arriving network traffic. In this approach any subsequent behaviour which does not match the learned 'normal' one is flagged as potential intrusion attempt.

The Misuse Detection category of IDS are based on existent library (or trained model) of attack signatures. All current activities are then matched against that knowledge base in order to identify any malicious behaviour. Most commonly these are rule-based models [10, 11] which often also employ a pattern-matching algorithm such as Boyer-Moore [12, 13].

Both misuse and anomaly detection have their positive and negative aspects. On the one hand, anomaly detection can be considered as flexible because it can detect new threats, requiring the creation or learning of a standard behaviour profile only. However, minimising the false-positives is still a significant challenge for this approach. Alternatively misuse detection, which can be considered as non-flexible because it requires pre-definition of known signatures or scenarios of attacks, but is often considered more reliable. An added advantage of misuse detection is that it also allows the classification of the intrusion attempt into different types and categories. Increasingly its main issue however is that it requires a manually encoded rule set in advance.

Port scanning is usually one of the initial step in any intrusion attempt. A variety of port-scanning detection approaches have been proposed in the literature, including probabilistic model, fuzzy logic, rule-based, with the most popular being a simple threshold-based - it is implemented in Snort [14, 15].

2   Model and Experimental Setup

The experiments were conducted using a rooted Android Nexus 7 tablet connected to the network via WiFi. The model was designed partially following the Common Intrusion Detection Framework (C.I.F.D) [4]: Event Generator (Data Gathering Module), Event Analyzer (Pre-processing, Pattern matching and Post-processing Modules), Response Unit (Alert Module) (figure 1).

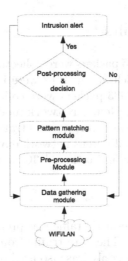

Fig. 1. Overall design of the model

Data Gathering Module is using TcpDump-arm to capture all network packets received to or transmitted from the device. The model presented here is not using the content of the packets but only the TCP and IP headers. The *Pre-processing* module extracts from the TcpDump output only the information required:

RT flag identifying received (IN) or transmitted (OUT) packet
SIP source address from IP packet header
DIP destination address from IP packet header
SP source port from TCP packet header
DP destination port from TCP packet header
FLG TCP header flags set
TST timestamp of the packet

Based on these features the following data is accumulated and sent to the *Pattern Matching* module:

RT   flag identifying IN/OUT packet
IPT   frequency of IN ports (ports-per-second for remote IP)
IPK   frequency of IN packets (packets-per-second for remote IP)
OPT   frequency of OUT ports (ports-per-second for remote IP)
OPK   frequency of OUT packets (packets-per-second for remote IP)
FLG   TCP header flags set for Decision tree model,
      or is flag set valid indicator for the Neural Network model

In the separate experiments, the Pattern Matching module was designed using a Decision Tree and a Cascade Correlation Neural Network [5]. Finally, the output of the Pattern matching module was passed to the Decision module which depending on whether a potential intrusion was identified could rise he Alert or continue normal processing.

3   Experimental Results

A total of approximately 10665 packets were collected for the development of the Desktop Decision tree, of which about 3112 were from port scans and the rest were normal traffic. About 49683 packets were collected for the development of the Android Decision tree and Neural Network models, of which approximately 21744 were port scans and the rest were normal traffic. The data from the mobile device was collected from four different network environments: a tethered wired connection, a WiFi connection and tethered 3G and 4G connections.

3.1   Decision Tree IDS

The data collected from the Desktop traffic was preprocessed and used to built a Decision tree model (figure 2). The data traffic from the Android device for the connection to the WiFi network only was also used to built a Decision tree model (figure 3). As can be seen from the initial root node of the trees and the further decision nodes, the frequency within which packets are received on the mobile platform is noticeably lower for the mobile device. If the Desktop Decision tree were to be transferred directly onto the mobile device it would be producing a considerable number of false positives in the rules defines in the the right branch (i.e. missing potential attacks).

The Decision tree of the mobile device was further tested on different speeds and types of scan. Table 1 shows the results of the port scan detection with the different scan timing options of Nmap. The Decision tree model was capable of detecting the normal and fast speed port scans but was struggling with the slower ones.

In a further experiment the Decision tree was manually extended to incorporate the flag sets from the TCP header in order to identify the type of scan it has detected. Table 2 shows the results in which all basic types of scans were

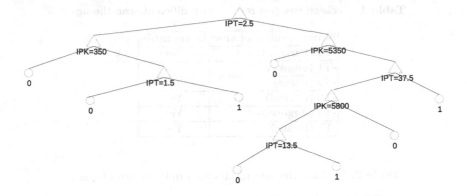

Fig. 2. Decision tree built for the desktop IDS

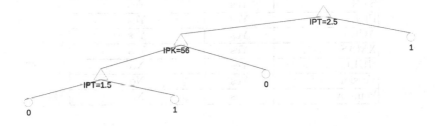

Fig. 3. Decision tree built for the mobile device IDS

correctly identified, however, the model was not able to cope with the more complex ones.

3.2   Android Neural Network IDS

Neural Networks provide a number of advantage in the development of IDS: in general Neural networks are relative easy to train and retrain compared to some statistical models and require fewer assumptions to be made in advance. Furthermore, once built a Neural network has relatively low running cost in live environment. Different models have been proposed for port scanning detection and intrusion detection in general, including Self-Organising Maps, Radial Basis Function, Random Neural Networks, Multi-layer Perceptron [16–21].

Cascade Correlation Neural Networks (CCNN) [5] allow the training of the network to new data without the requirement to retrain the whole network with the original data as well. When new events are detected, new hidden nodes are added to the network and only those are trained with the newly collected data - allowing for a run time adaptive and scalable system.

Initially the Neural network was trained only with the data from the mobile device connected to the tethered LAN and the WiFi connection. For all experiments the available data was split into 60% training, 20% validation and 20%

Table 1. Decision tree test results with different scan timings

Nmap timing option	Detectable
-T0 (Paranoid)	No
-T1 (Sneaky)	No
-T2 (Polite)	Yes
-T3 (Normal)	Yes
-T4 (Aggressive)	Yes
-T5 (Insane)	Yes

Table 2. Decision tree test results with different scan types

Nmap scan type	Detectable	Scan type recognised
TCP	Yes	Yes
SYN	Yes	Yes
FIN	Yes	Yes
ACK	Yes	Yes
XMAS	Yes	Yes
NULL	Yes	No
Version	Yes	No
Window	Yes	No

test data sets and trained with Levenberg-Marquardt Error Backpropagation. The trained network had 14 hidden neurones and the test result matched those of the Decision tree (table 3). At the next stage, the Neural network was trained on the data collected from the 3G connection and then on the data collected from the 4G connection. Those training sessions added another 7 and 4 hidden neurones to the network respectively. As can be seen from the result in table 3, the model is now capable of detecting all port scans on the various Nmap timing options, including the very slow ones.

Table 3. Neural Network test results with different scan timings

Nmap timing option	Detectable after LAN and WiFi training	Detectable after further 3G and 4G training
-T0 (Paranoid)	No	Yes
-T1 (Sneaky)	No	Yes
-T2 (Polite)	Yes	Yes
-T3 (Normal)	Yes	Yes
-T4 (Aggressive)	Yes	Yes
-T5 (Insane)	Yes	Yes

4 Conclusions

The modelling and experimental results presented here show that a Neural network model can be trained on the task of detecting active port scans on mobile devices with performance matching that of the more traditional rule-based systems. However, the Neural network based approach provides a more scalable and adaptive solution which is capable of learning in new environments while being computation cost effective - as the devices move into new networks and environments, new data can be collected and the network can be trained on that data simply by adding and training the new nodes. The model presented here is the first application of a Cascade Correlation Neural Network to the detection of port scans on mobile devices. The neural network model was coded and installed on a Nexus 7 tablet. The mobile device was then taken to a few public ares with available WiFi connections and on two occasions an actual port scan was detected. Network traffic on mobile devices can be a source of significant amounts of data, making it possible to extend the work by implementing Deep Learning models - this is a direction of future work we are currently exploring.

References

1. Khan, S., Nauman, M., Othman, A.T., Musa, S.: How secure is your smartphone: An analysis of smartphone security mechanisms. In: Proceedings of the 2012 International Conference on Cyber Security, Cyber Warfare and Digital Forensic (CyberSec), June 26-28, pp. 76–81 (2012)
2. Zaman, S., Karray, F.: TCP/IP Model and Intrusion Detection Systems. In: Proceedings of the International Conference on Advanced Information Networking and Applications Workshops, Bradford, United Kingdom, May 26-29, pp. 90–96 (2009)
3. Kou, X., Wen, Q.: Intrusion detection model based on Android. In: Proceedings of the 4th IEEE International Conference on Broadband Network and Multimedia Technology (IC-BNMT), pp. 624–628 (2011)
4. Ghorbanian, M., Shanmugam, B., Narayanasamy, G., Idrids, N.: Signature-Based Hybrid Intrusion detection system (HIDS) for Android devices Business Engineering and Industrial Applications Colloquium (BEIAC), April 7-9, pp. 827–831. IEEE (2013)
5. Fahlman, S.E., Lebiere, C.: The cascade-correlation learning architecture. In: Touretzky, D.S. (ed.) Advances in Neural Information Processing Systems 2, pp. 524–532. Morgan Kaufmann Publishers Inc., San Francisco (1990)
6. Govindarajan, M., Chandrasekaran, R.M.: Intrusion detection using k-Nearest Neighbor. In: Proceedings of the First International Conference on Advanced Computing ICAC, December 13-15, pp. 13–20 (2009)
7. Jie, Y., Chen, X., Xiang, X., Wan, W.: HIDS-DT: An Effective Hybrid Intrusion Detection System Based on Decision Tree International Conference on Communications and Mobile Computing, April 12-14, pp. 70–75 (2010)
8. Gates, C., Taylor, C.: Challenging the anomaly detection paradigm: a provocative discussion. In: Proceedings of the Workshop on New Security Paradigms (NSPW 2006), New York, USA, pp. 21–29 (2006)
9. Denning, D.E.: An Intrusion-Detection Model. In: Proceedings of the IEEE Symposium on Security and Privacy, pp. 118–133 (1986)

10. Mitchell, R., Chen, I.-R.: "Behavior-Rule Based Intrusion Detection Systems for Safety Critical Smart Grid Applications. IEEE Transactions on Smart Grid 4(3), 1254 (2013)
11. Yang, Y., McLaughlin, K., Littler, T., Sezer, S., Wang, H.F.: Rule-based intrusion detection system for SCADA networks. In: 2nd IET Renewable Power Generation Conference (RPG 2013), September 9-11, pp. 1–4 (2013)
12. Boyer, R.S., Moore, J.S.: A Fast String Searching Algorithm. Comm. ACM 20(10), 762–772 (1977)
13. Antonatos, S., Polychronakis, M., Akritidis, P., Anagnostakis, K.G., Markatos, Y.E.P.: Fast and Memory-Efficient Pattern Matching for Intrusion Detection. In: Proceedings 20th IFIP International Information Security Conference SEC (2005)
14. Bhuyan, M., Bhattacharyya, D.K., Kalita, J.K.: Surveying Port Scans and Their Detection Methodologies. Computer Journal ACM 54, 1565–1581 (2011)
15. Dabbagh, M., Ghandour, A.J., Fawaz, K., Hajj, W.E., Hajj, H.: Slow port scanning detection. In: Proceedings of the 7th International Conference on Information Assurance and Security (IAS), December 5-8, pp. 228–233 (2011)
16. Wang, G., Hao, J., Ma, J., Huang, L.: A new approach to intrusion detection using Artificial Neural Networks and fuzzy clustering. Expert Systems with Applications 37(9), 6225–6232 (2010)
17. Nazir, A.: A comparative study of Cascaded Forward Back Propagation and Hybrid SOFM-CFBP Neural Networks based Intrusion Detection Systems. International Journal of Scientific and Engineering Research 4(6) (2013)
18. Basu, R., Cunningham, R.K., Webster, S.E., Lippmann, R.P.: Detecting low-profile probes and novel denial-of-service attacks. In: Proceedings of IWIAS 2001, West Point, New York, USA, pp. 5–10. IEEE Computer Society (June 2001)
19. Oke, G., Loukas, G., Gelenbe, E.: Detecting denial of service attacks with bayesian classifiers and the random neural network. In: Proceedings of FUZZ- IEEE 2007, pp. 1964–1969. IEEE, USA (2007)
20. Fisch, D., Hofmann, A., Sick, B.: On the versatility of radial basis function neural networks: A case study in the field of intrusion detection. Information Sciences 180(12), 2421–2439 (2010)
21. Kalpana, Y., Purushothaman, S., Rajeswari, R.: Implementation of Echo State Neural Network and Radial Basis Function Network for Intrusion Detection. Data Mining and Knowledge Engineering 5(9), 366–373 (2013)

Categorization and Construction of Rule Based Systems

Han Liu[1], Alexander Gegov[1], and Frederic Stahl[2]

[1] School of Computing, University of Portsmouth, Buckingham Building, Lion Terrace,
PO1 3HE, Portsmouth, United Kingdom
{Han.Liu,Alexander.Gegov}@port.ac.uk
[2] School of Systems Engineering, University of Reading, P.O. Box 225 Whiteknights, Reading,
RG6 6AY, United Kingdom
F.T.Stahl@reading.ac.uk

Abstract. Expert systems have been increasingly popular for commercial importance. A rule based system is a special type of an expert system, which consists of a set of 'if-then' rules and can be applied as a decision support system in many areas such as healthcare, transportation and security. Rule based systems can be constructed based on both expert knowledge and data. This paper aims to introduce the theory of rule based systems especially on categorization and construction of such systems from a conceptual point of view. This paper also introduces rule based systems for classification tasks in detail.

Keywords: Data Mining, Machine Learning, Rule Based Systems, Rule Based Classification, if-then Rules.

1   Introduction

The development of rule based systems began in the 1960's but became popular in the 1970's and 1980's [1]. A rule based system typically consists of a set of if-then rules, which can serve many purposes such as decision support or predictive decision making in real applications. One of the main concerns in this area is the construction of such systems which could be based on both expert knowledge and data. Thus the construction techniques can be divided into two categories: knowledge based construction and data based construction. This paper introduces the theoretical aspects of categorization and construction of rule based systems as well as the use for classification tasks. The purpose is to explore the research direction in context as well as combine the authors' previous work together to make an evolution from specialization to generalization for the theoretical concepts.

The rest of this paper is organized as follows: Section 2 introduces the categorization of rule based systems according to some special characteristics; Section 3 introduces two main categories of construction of rule based systems: knowledge based construction and data based construction. A special type of rule based systems used for classification tasks is introduced in detail in Section 4. The potential of this approach is also specified in a healthcare case study in Section 5 to

V. Mladenov et al. (Eds.): EANN 2014, CCIS 459, pp. 183–194, 2014.

demonstrate the value and impact of the approach. The summary of completed work and further directions of research in this area are highlighted further in Section 6.

2   Categorization of Rule Based Systems

Rule based systems can be categorized in the following aspects: *number of inputs and outputs, type of input and output values, type of structure, type of logic, type of rule bases, number of machine learners and type of computing environment.*

For rule based systems, both inputs and outputs could be single or multiple. From this point of view, rule based systems can be divided into four types [2] with respect to number of inputs and outputs: *single-input-single-output, multiple-input-single-output, single-input-multiple-output, and multiple-input-multiple-output.* All the four types above can fit the characteristics of association rules. This is because association rules reflect the relationship between attributes. An association rule may have a single or multiple rule terms in both antecedent (left hand side) and consequent (right hand side) of the rule. Thus the categorization based on number of inputs and outputs is very relevant to fulfill the distinction of association rules.

However, association rules include two special types: classification rules and regression rules, depending on type of output values. Both classification rules and regression rules may have a single or multiple rule terms in antecedent but can only have a single term in the consequent. The difference between classification rules and regression rules is that the output values of classification rules must be discrete while those of regression rules must be continuous. Thus both classification rules and regression rules fit the characteristics of 'single-input-single-output' or 'multiple-input-single-output' and are seen as special type of association rules. As the basis of above description, rule based systems can also be categorized into three types with respects to both number of inputs and outputs and type of input and output values: *rule based classification systems, rule based regression systems and rule based association systems.*

In machine learning, classification rules can be generated in two approaches: divide and conquer [3] and separate and conquer [4]. The former method is generating rules directly in the form of a decision tree, whereas the latter method produces a list of 'if-then' rules. An alternative structure called Rule Based Networks represents rules in the form of networks. With respect to structure, rule based systems can thus be divided into three types: *treed rule based systems, listed rule based systems and networked rule based systems.*

Construction of rule based systems is based on special type of logic such as Boolean logic, fuzzy logic and probabilistic logic. From this point of view, rule based systems can also be divided into the following types: *deterministic rule based systems, probabilistic rule based systems and fuzzy rule based systems.*

As rule based systems can also be in the context of rule bases including single rule bases, chained rule bases and modular rule bases. From this point of view, rule based systems can also be divided into the three types: *standard rule based systems, hierarchical rule based systems and networked rule based systems.*

In machine learning context, a single algorithm could be applied to a single data set for training a single learner. It can also be applied to multiple samples of a data set by ensemble learning techniques for construction of an ensemble learner which consists of a group of single learners. In addition, there could also be a combination of multiple algorithms involved in machine learning tasks. From this point of view, rule based systems can be divided into two types according to the number of machine learners constructed: *single rule based systems and ensemble rule based systems*.

In practice, an ensemble learning task could be done in parallel, distributed way or a mobile device according to the specific computing environments. Therefore, rule based systems can also be divided into the following three types: *parallel rule based systems, distributed rule based systems and mobile rule based systems*.

3   Construction of Rule Based Systems

As mentioned in Section 1, the construction of rule based systems can be based on both expert knowledge and data. This section introduces and discusses two special types of construction: *knowledge based construction* and *data based construction*.

3.1   Knowledge Based Approach

Knowledge based construction follows a traditional engineering approach, which is in general domain dependent. It is necessary to have knowledge or requirements acquired from experts at first and then to identify the relationships between attributes (features). Modelling, which is the most important step, is further to be executed in order to build a set of rules. Once the modelling is complete, then simulation is started to check the model towards fulfillment of systematic complexity such as model accuracy and efficiency. Finally, statistical analysis is undertaken in order to validate whether the model is reliable and efficient in application.

3.2   Data Based Approach

Data based construction follows a machine learning approach, which is in general domain independent. Machine learning techniques can be subdivided into two types: supervised learning and unsupervised learning. Supervised learning means learning with a teacher. This is because all instances from a data set are labelled. The aim of this type of learning is to predict attribute values for unknown instances by using the known data instances [5]. The predicted value of an attribute may be either discrete or continuous. Therefore, supervised learning could be involved in both classification and regression tasks for categorical prediction and numerical prediction respectively. On the other hand, unsupervised learning means learning without a teacher. This is because all instances from a data set are unlabeled. The aim of this type of learning is to find previously unknown patterns from data sets. It includes association, which aims to find relationships among attributes with regards to their values [5], and clustering, which aims to find a group of objects that are similar from data sets [5].

As mentioned in Section 1, rule based systems can be used for construction of classification, regression and association systems. In general, all the three types of rule based systems can be constructed with the following steps: Data collection->Data pre-processing->Learning from data->Testing. However, there are different requirements in different learning tasks. In other words, in order to build a high quality model by using machine learning techniques, it is important to find algorithms which are suitable to the chosen data sets with respects to the characteristics of data. From this point of view, data preprocessing may be not necessary if the chosen algorithms are good fits. In addition, different type of dimensionality reduction techniques (such as feature selection), a type of data preprocessing, may be required for different tasks. If it is a classification or regression task, supervised feature selection techniques may be required in general. Otherwise unsupervised feature selection techniques may be suitable. The step for learning from data mentioned above may also need to be broken down in some special cases. For example, it may be required to simplify rules in classification tasks or to reduce the number of rules in association tasks. A specific construction for rule based classification systems is further introduced in more detail in Section 4.

3.3   Discussion

In this paper, the authors aim to motivate the use of data based approach instead of knowledge based approach for construction of complex rule based systems. The main reason is that expert knowledge may be incomplete or inaccurate; some of experts' points of view may be biased; engineers may misunderstand requirements or have technical designs with defects. In other words, with regards to solving problems with high complexity, both domain experts and engineers are difficult to have all possible cases considered or to have perfect technical designs. Once a failure arises with such a system, experts or engineers may have to find the problem and fix it by reanalyzing or redesigning. However, the real world has been filled with Big Data. Some previously unknown information or knowledge may be discovered from data. Data may potentially be used as supporting evidence to reflect some useful and important pattern by using modeling techniques. More importantly, the model could be revised automatically as the update of database in real time if data based modeling technique is used. Therefore, data based approach may be more suitable than knowledge based approach for construction of complex systems. The rest of the paper will focus on discussion in the machine learning context.

4   Rule Based Classification Systems

In general, a unified framework for the construction of predictive rule based systems, comprises three basic procedures, the generation of rules, the simplification of rules and the rule representation.This section describes the essence of the three operations and introduces some methods and techniques which are involved in the operations. The methods and techniques are also discussed comparatively in order to highlight some important aspects in choosing methods or techniques for the fulfillment of each of the three operations.

4.1   Rule Generation

As mentioned in Section 2, the methods for generation of classification rules can be categorized into the 'divide and conquer' and the 'separate and conquer' approaches. Examples for 'divide and conquer' comprise ID3 [3], C4.5 and C5.0. Examples for 'separate and conquer' comprise Prism [7] and PrismTCS [8].

Divide and conquer is a recursive approach as the generation of rules is to select an attribute to split on and then to recursively repeat the process for each branch covering a subset of the training set as illustrated in Fig.1. However, this approach has a principal drawback, the replicated sub-tree problem pointed out in [7] and illustrated in Fig.2. It can be seen from Fig.2 that the four sub-trees which all have node C as root are identical. This is an unnecessary redundancy in the decision tree as illustrated in Fig.2.

IF all cases in the training set belong to the same class
THEN return the value of the class
ELSE

    (a)   Select the attribute A to split on*
    (b)   Sort the instances in the training set into non-empty subsets, one for each value of attribute A
    (c)   Return a tree with one branch for each subset, each branch having a descendant sub-tree or a class value produced by applying the algorithm recursively for each subset in turn.

*When selecting attributes at step (a) the same attribute must not be selected more than once in any branch.

Fig. 1. TDIDT Tree Generation algorithm [5]

As the problem arises with the rule generation approach, the separate and conquer approach is motivated to generate if-then rules directly and iteratively from training instances. Prism is a method that follows the 'separate and conquer' approach and is illustrated in its original form in Fig.3. Bramer developed a modified version of Prism called PrismTCS. The motivation is to increase computational efficiency because original Prism is computationally expensive. The expensive computation is resulted from frequent deletion of instances during rule generation and restoring the training data to its initial size for rule generation for next class [12]. PrismTCS always chooses the minority class as target class. Thus PrismTCS induces rules in the order of their importance without the restoring the data to its original size (in between the induction of different rules) [8, 9, 10]. PrismTCS has shown to produce classification rules much faster, but also of a similar level of predictive accuracy compared with original Prism [8, 9, 13]. However, the authors have recently pointed out some limitations of Prism algorithm in [6, 14] regarding Prism's way of dealing with clashes, underfitting of the concept in the training data and its computational efficiency.

With respects to clashes, it indicates that Prism may generate a number of rules, each of which covers a clash set. A clash set contains instances that belong to different classifications but cannot be separated further. According to Bramer's Inducer software implementation for clash handling, Prism prefers to discard a rule instead of assigning it to the majority class. It may result in underfitting of the training set if a large number of rules get discarded. For original Prism, this case may result in a large number of instances remaining unclassified as there is no default rule available and the rules that cover the instances get discarded. For PrismTCS, this case may make a default rule give wrong classifications to the instances covered by discarded rules. This is because the default rule is supposed to cover only the instances that belong to the majority class, but unfortunately some rules that cover the other instances got discarded.

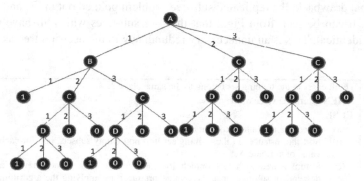

Fig. 2. Cendrowska's replicated subtree example [6, 20]

With respects to computational efficiency, as mentioned above, Prism prefers to discard a rule if a clash occurs. That indicates that the algorithm takes time to generate a rule which is eventually discarded in some cases. It is equivalent to doing nothing and results in unnecessary computational costs.

Execute the following steps for each classification (*class= i*) in turn and on the original training data *S*:
1. *S'=S*.
2. Remove all instances from *S'* that are covered from the rules induced so far. If *S'* is empty then stop inducing further rules
3. Calculate the conditional probability from *S'* for *class=i* for each *attribute-value pair*.
4. Select the *attribute-value pair* that covers *class= i* with the highest probability and remove all instances from *S'* that comprise the selected *attribute-value pair*
5. Repeat 3 and 4 until a subset is reached that only covers instances of *class= i* in S'. The induced rule is then the conjunction of all the *attribute-value pairs* selected.
Repeat 1-5 until all instances of *class i* have been removed
*For each rule, no one attribute can be selected twice during rule generation

Fig. 3. Basic Prism algorithm [5]

The authors have recently developed another rule generation method called "Information Entropy Based Rule Generation" (IEBRG) which also follows separate and conquer approach and is illustrated in Fig.4. However, it uses "from cause to effect" approach whereas Prism uses "from effect to cause" approach. The main focus of IEBRG is on minimizing the uncertainty that exists in the subset no matter what the target class is. A popular measure of uncertainty is information entropy introduced by Shannon in [15]. One of the advantages of IEBRG compared with Prism can be seen from an example with reference to the lens 24 dataset reconstructed by Bramer in [5]. The dataset indicates that p $(class=3/tears=1)$ $=1$. The first rule generated could be "if tears=1 then class=3".This implies that "tears=1" is only relevant for predicting class 3. IEBRG can capture this information by the conditional entropy E $(tears=1)$ $=0$. However, this is actually unknown prior to the rule induction by Prism algorithm. The PrismTCS would assign *class 1* as target class to the first rule being generated (as *class 1* is the minority class). Original Prism may also select *class 1* as the index of the class is smaller. However, according to [8] the first rule generated by original Prism is "if astig=2 and tears=2 and age=1 then class=1". It indicates that the computational cost is slightly higher than expected and so the rule has a higher complexity. In some cases, the Prism algorithm may be even generating incomplete rules, covering a clash set, especially if the target class is not a good fit to the attribute-value pairs in the training data. The rule may be discarded resulting in underfitting and unnecessary computational cost.

---

1. Calculate the conditional entropy of each attribute-value pair in the current subset
2. Select the attribute-value pair with the smallest entropy to be spilt on, i.e. remove all other instances that do not comprise the attribute-value pair.
3. Repeat step 1 and 2 until the current subset contains only instances of one class (the entropy of the resulting subset is zero).
4. Remove all instances covered by this rule.

Repeat 1-4 until there are no instances remaining in the training set.

* For each rule, no one attribute can be selected more than once during generation.

Fig. 4. IEBRG algorithm

In comparison with the Prism algorithm family, IEBRG may need significantly less computational effort. In contrast to Prism, the IEBRG algorithm deals with clashes by assigning a majority class to the rule. This may potentially reduce the underfitting of the rule set and thus reduce the number of unclassified instances. However, there is potential that the number of misclassified instances increases. Yet, IEBRG is potentially better in avoiding clashes compared with Prism.

4.2   Rule Simplification

Rule simplification is necessary in some cases. The reason is the principal problem of rule based classifiers to overfit on the training data [17]. When a large data set is used for training, this may lead to the induction of a very large number of complex rules. This will lower both the predictive accuracy and the computational efficiency. This has motivated the development of pruning methods for rule simplification with respect to the reduction of overfitting. Pruning methods can be subdivided into two categories- pre-pruning [5] and post-pruning [5]. The former prunes rules during rule generation and the latter generates a whole rule set and then discards a number of rules and rule terms, by means of using statistical (or other) tests [17]. There is a family of pruning algorithms for Prism algorithms based on the J-measure [18], an information theoretic means to compute the theoretical information content of a rule. This is based on the hypothesis [19] that, if a rule has high information content (value of J-measure, or also called J-value), it is also prone to have a high classification accuracy. Two existing J-measure based pruning algorithms are J-pruning [17] and Jmax-pruning [9, 10]. They have been successfully applied on different versions of Prism algorithms for reducing overfitting. When a rule is being generated, the J-value may go up or go down after specialising the rule by appending an additional term. Both pruning algorithms expect to find the global maximum of J-value for the rule. Each rule is assigned a complexity degree which is the number of terms. The increase of complexity degree may make the J-value of this rule go up or down. The pruning algorithms are aimed at finding the complexity degree corresponding to the global maximum of J-value as illustrated in Fig. 5 using a fictitious example. Both pruning methods above employ different strategies to search for the global maximum of the J-value. J-pruning monitors the change of the J-value when appending rule terms and stops once the J-value goes down. In contrast, Jmax-pruning induces the rule fully until complexity degree X_3 (regarding Fig.4), but monitors and records the so far highest J-value when appending rule terms. In the example in Fig.5, J-pruning would stop inducing rule terms when reaching complexity degree X_1 but Jmax-pruning would stop when reaching complexity degree X_3 and then reduce the complexity degree to X_2 by removing rule terms between X_3 and X_2.

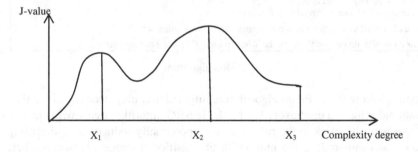

Fig. 5. Relationship between complexity degree and J-value

The authors have recently developed an alternative J-measure based pruning algorithm called Jmid-pruning [20] in order to overcome the limitations mentioned above. This algorithm not only monitors and records the highest J-value observed but also measures the Jmax value that may be achieved by adding further rule terms.

In comparison with Jmax-pruning, Jmid-pruning also always finds the global maximum but it is in theory computationally more efficient in some cases. An example [11] is considered that a rule could be generated using the lense24 dataset:

If tears=2 and astig=1 and age=3 and specRx =1 then class= 3;

As the rule is being specialized by appending the four terms subsequently, the corresponding values of J and Jmax change in the pattern as follows:

If tears=2 then class=3; (J=0.210, Jmax=0.531)

If tears=2 and astig=1 then class=3; (J=0.161, Jmax=0.295)

If tears=2 and astig=1 and age=3 then class=3; (J=0.004, Jmax=0.059)

If tears=2 and astig=1 and age=3 and specRx =1 then class= 3; (J=0.028, Jmax=0.028)

In the example above all three pruning algorithms would generate the same rule: *if tears=2 then class=3*. The reason is that the highest J-value is computed right after the first rule term was added (tears=2). However, with regard to computational efficiency, J-pruning is the fastest and stops right after the second term (astig=1) is generated. Jmid-pruning is faster than Jmax-pruning. This is because Jmax-pruning stops when the rule is complete and cuts it back to 'if tears=2 then class=3' but Jmid-pruning stops the generation after the third term is generated as the Jmax-value is below the so far highest J-value.

4.3   Rule Representation

Rule representation aims to represent a rule set in a suitable structure to achieve more efficient prediction. As mentioned in Section 2, the existing rule representations include decision tree and linear list. The former is a representation that automatically represents classification rules induced using the 'divide and conquer' method. The latter automatically represents rules generated by the 'separate and conquer' method. However, the decision tree representation has been criticised by Cendrowska and identified as a major reason for overfitting [7]. It is also pointed out in [16] that in the worst case it needs to go through the entire tree for extracting a classification. It undoubtedly increases the computational costs and thus is a major drawback, hence the motivation for using 'if-then' that can be represented in a linear list structure. However, prediction on test instances by the list representation is done in linear time while the number of rule terms in the rule set is the input size n. It indicates it may have to go through the whole rule set in the worst case in order to find the first rule firing. This may result in huge computational costs in prediction stage when a rule set is very complex. Therefore, the authors have recently developed a new representation called Rule Based Classification Networks [6] which performs logarithmic time.

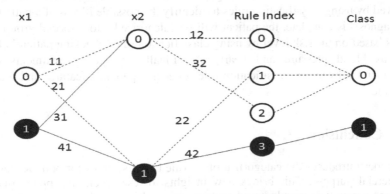

Fig. 6. Rule Based Classification Networks [6]

The networked representation is illustrated in Fig.6 to represent a rule set based on Boolean logic. The rule set has two input attributes (x1 and x2) and the class value is 1 if and only if both variables get input value of 1. In this representation, the terms: x1=1 and x2=1, are the two inputs for testing. Thus, in both 'x1' and 'x2' layers, node labeled 1 becomes black and node labeled 0 becomes white. This is because each node in layer x1 represents a value of attribute x1 and equivalent in layer x2. The two digits with which the connections between layer x1 and x2 are labeled represent the index of the rule and rule term respectively, i.e. the two digits '11' state that this is the first term of the first rule. It can also be easily seen that this particular term is 'x1=0'. However, as the value of x1 is 1, the connection is not satisfied and thus becomes dot. The connections '31' and '41' are both solid because condition 'x1=1' is met. The same principle applies to the connections between layers 'x2' and 'Rule Index'. The connections '31', '41' and '42' become solid as the inputs are x1=1 and x2=1; and this thus results in that node 3 becomes black in the 'Rule Index' layer and the output is 1 in the layer 'Class'.

5   Applications in Healthcare

As mentioned in Section 3, rule based classification systems constructed by machine learning approach are domain independent and thus can be applied in many areas. For example, as mentioned in [21], Inductive learning algorithms are domain independent and can be involved in any classification or pattern recognition tasks. Some successful applications listed in [21] include lymphography, prognosis of breast cancer recurrence, location of primary tumour and thyroid problem diagnosis in medicine [22, 23, 24].

The authors have recently developed a healthcare process modeling approach which could be implemented in the following procedures using classification rules:

- diagnosis of illness
- patient classification
- treatment recommendation

It detail, patients need to have diagnosis with regards to the illness. This could be achieved by using classification rules to identify the possible illness of a patient. Once the diagnosis is complete, the patient will be categorized into a special group for this illness based on the risk level by using classification rules checking patient's features such as blood pressure and heart rate. Finally, a list of treatments can be recommended by using classification rules checking patient's features and finding all fired rules.

6   Conclusion

This paper introduces the categorization of rule based systems for both academic and commercial purpose. This brings new insights to researchers and practitioners and positions a new type of rule based systems for applications. It also motivates the use of data based approach in the context of machine learning for construction of complex systems instead of knowledge based approach. The importance of the data based approach is also highlighted. A special type of rule based system used for classification tasks is introduced in detail explaining the construction framework and reflecting some important aspects of choosing methods or techniques for rule generation, simplification and representation operations. This is in order to explore the significance of data based approach in depth. In addition, a healthcare case study is also specified to demonstrate the value and impact of the approach. The construction framework could be extended to include multiple rule based classification systems as a whole by means of a system of systems by adopting an ensemble learning approach. Such an extended framework for constructing ensemble rule based systems will be validated. The studies will also be extended towards fulfilment for construction of rule based systems for regression and association tasks.

References

1. Partridge, D., Hussain, K.M.: Knowledge Based Information Systems. Mc-Graw Hill (1994)
2. Gegov, A.: Fuzzy Networks for Complex Systems: A Modular Rule Base Approach. Springer, Berlin (2010)
3. Quinlan, J.R.: C4.5: Programs for Machine Learning. Morgan Kaufman (1993)
4. Michalski, R.S.: On the Quasi-Minimal solution of the general covering problem. In: Proceedings of the Fifth International Symposium on Information Processing, Bled, Yugoslavia, pp. 125–128 (1969)
5. Bramer, M.A.: Principles of Data Mining. Springer, London (2007)
6. Liu, H., Gegov, A., Stahl, F.: Unified Framework for Construction of Rule Based Classification Systems. In: Pedrycz, W., Chen, S.M. (eds.) Springer, Berlin (in press)
7. Cendrowska, J.: PRISM: An Algorithm for Inducing Modular Rules. International Journal of Man-Machine Studies 27, 349–370 (1987)
8. Bramer, M.A.: Automatic Induction of Classification Rules from Examples using N-Prism. Research and Development in Intelligent Systems, vol. XVI, pp. 99–121. Springer, Cambridge (2000)
9. Stahl, F., Bramer, M.A.: Jmax-pruning: A Facility for the Information Theoretic Pruning of Modular Classification Rules. Knowledge-Based Systems 29, 12–19 (2012)
10. Stahl, F., Bramer, M.A.: Induction of Modular Classification Rules: using Jmax-pruning. In: Thirtieth SGAI International Conference on Innovative Techniques and Applications of Artificial Intelligence, pp. 14–16. Springer, Heidelberg (2011)
11. Bramer, M.A.: Inducer: a Public Domain Workbench for Data Mining. International Journal of Systems Science 36(14), 909–919 (2005)
12. Stahl, F., Bramer, M.A.: Computationally Efficient Induction of Classification Rules with the PMCRI and J-PMCRI Frameworks. Knowledge-Based Systems 35, 49–63 (2012)
13. Bramer, M.A.: An Information-theoretic Approach to the Pre-pruning of Classification Rules. In: Musen, B.N., Studer, R. (eds.) Intelligent Information Processing, pp. 201–212. Kluwer (2002)
14. Liu, H., Gegov, A.: Induction of Modular Classification Rules by Information Entropy Based Rule Generation. In: Sgurev, V., Yager, R., Kacprzyk, J. (eds.) Innovative Issues in Intelligent Systems. Springer (in press)

15. Shannon, C.: A Mathematical Theory of Communication. Bell System Technical Journal 27(3), 379–423 (1948)
16. Deng, X.: A Covering-based Algorithm for Classification: PRISM. CS831: Knowledge Discover in Databases (2012)
17. Bramer, M.A.: Using J-Pruning to Reduce Overfitting of Classification Rules in Noisy Domains. In: Hameurlain, A., Cicchetti, R., Traunmüller, R. (eds.) DEXA 2002. LNCS, vol. 2453, p. 433. Springer, Heidelberg (2002)
18. Smyth, P., Goodman, R.M.: Rule Induction Using Information Theory. In: Piatetsky-Shapiro, G., Frawley, W.J. (eds.) Knowledge Discovery in Databases, pp. 159–176. AAAI Press (1991)
19. Bramer, M.A.: Using J-Pruning to Reduce Overfitting in Classification Trees. In: Research and Development in Intelligent Systems XVIII, pp. 25–38. Springer (2002)
20. Liu, H., Gegov, A., Stahl, F.: J-measure Based Hybrid Pruning for Complexity Reduction in Classification Rules. WSEAS Transaction on Systems 12(9), 433–446 (2013)
21. Aksoy, M.S.: A Review of Rules Families of Algorithms. Mathematical and Computational Applications 13(1), 51–60 (2008)
22. Quinlan, J.R.: Induction, Knowledge and Expert Systems. In: Gero, J.S., Stanton, R. (eds.) Artificial Intelligence Developments and Applications, Amsterdam, North Holland, pp. 253–271 (1988)
23. Michalski, R.S., et al.: The Multi-purpose Incremental Learning System AQ15 and Its Testing Application to Three Medical Domains. In: Proc. National Conf. on AI, Philadelphia, PA, pp. 1041–1044 (August 1996)
24. Quinlan, J.R.: Inductive Knowledge Acquisition: a Case Study. In: Quinlan, J.R. (ed.) Applications of Expert Systems, Quinlan, J, pp. 157–173. Turing Institute Press (1987)

Tiling of Satellite Images to Capture an Island Object

Ahmet Sayar[1], Süleyman Eken[1], and Umit Mert[2]

[1] Computer Engineering Department, Kocaeli University,
41380 Izmit, Turkey
{ahmet.sayar,suleyman.eken}@kocaeli.edu.tr
[2] Information Technologies Institute,
The Scientific and Technological Research Council of Turkey, Gebze, Kocaeli, Turkey
umit.mert@tubitak.gov.tr

Abstract. This study proposes a novel tiling approach to capture an image of an entire object. Multi-spectral and multi-temporal satellite images are obtained a priori, and these individual image pieces can then be joined together at a later date to form an image of the entire object. The effectiveness of the proposed technique has been studied by tiling partially overlapping satellite mosaic images of the Island of Cyprus. The images were captured by the recently-launched LandSat-8 satellite.

Keywords: Satellite image tiling, image mosaicking, LandSat-8, lighten method.

1 Introduction

Processing satellite images is harder than processing any other images. It is even harder when it comes to image stitching or registration. There are a number of reasons for this. (1) Bad weather and atmospheric conditions such as clouds, fog and smoke affect the sensors and prevent them from acquiring measurements accurately. This also affects information extraction: It is hard to define objects with their precise borders and edges. (2) Satellite images are naturally poor in color variations. In low-resolution images, the colors are mostly greenish and brownish. This is related to the previous poor-definition group. (3) Satellite images from different sensors usually have different spatial resolution. (4) They are called multi-spectral images; images have different spectral characteristics, so that contrast information is different for the same imaged object. (5) They are mostly captured by the sensors at different time intervals. This is called spatiotemporal differences; and it affects the success of matching process in image stitching and registration. Issues 1 and 2 are general problems in satellite image processing. However, issues 3, 4 and 5 are especially related to the challenges faced when performing image stitching and registration.

In the image registration process, there are two major and important issues that need to be solved together. One is accuracy and another is efficiency. When you want to make the process more accurate, then you pay for the efficiency, and vice versa. Given the remotely captured large size images, the important thing is to reduce the

V. Mladenov et al. (Eds.): EANN 2014, CCIS 459, pp. 195–204, 2014.

computational time which is required to execute each of these steps while keeping precise alignment. It is therefore crucial that the registration process produces an image that is visually and numerically accurate.

In this paper, we are trying to find a way to get over aforementioned problems in image registrations in remote sensing domain by considering domain specific natures and characteristics of satellites and satellite images. We are proposing very fast and efficient registration method. But, in this approach there might be some faults in terms of numerical accuracy. The proposed framework and technique can be used efficiently in applications in which the performance and visually correctness has the highest precedence than the numerically correctness. Numerical correctness is called as exact pixel to (lat, long) address matching. In some real-time and internet based distributed applications accuracy can be tolerable to some extent for the sake of high performance.

After presenting the registration algorithm we increase the quality of the outcome image with a smoothing technique. In the mosaic result, the overlapping region shows the important brightness difference with the residual of the mosaic. We will apply a 'blending technique' to rarefy this effect. These methods are used to flat the overlapping area.

The study presented in this paper is based on mosaicking of the images obtained from an UAV (Unmanned Aerial Vehicle). UAV is referred to as a remotely piloted aircraft, when it flies in the sky to capture images remotely, and these are serially shot by digital camera installed on UAV. There are two methods used for image matching i.e. Rough Matching and Fine Matching. In the rough matching technique, firstly it determines the overlapping areas approximately and applies stitching.. This approximation is based on the mathematical calculations in which speed of UAV and some other parameters related to camera installed at UAV. In fine matching techniques, pixel based high cost feature detection algorithms are used. Techniques are mostly based on feature extraction and feature matching. To achieve a successful result, objects in the images need to be clearly determined. These approaches are affected by the problems presented in the beginning of this chapter.

The proposed tiling architecture is a kind of rough matching but also utilizes fine matching approaches to some extent. When coordinate values and all other required metadata about the satellite images are known and fed into the system, we can achieve much more efficient and successful results. In the proposed architecture, we handle the image stitching problems stemming from the spatiotemporal differences of satellite images. The architecture is also based on geometrical and coordinate based stitching by utilizing coordinate reference systems on which the satellite images are created.

The remainder of this paper is organized as follows. Section 2 gives relevant works about image stitching. Section 3 explains the proposed technique for registering the remote sensing satellite images. Section 4 presents the results of the experiment obtained by applying the technique on the Island of Cyprus mosaic image tiles. Section 5 concludes the paper.

2   Related Works

The methods used for image registration can be grouped according to the different perspectives and information used for registration. Mostly, they are grouped into two categories: feature-based and area-based methods [1]. In area-based methods, prominent features in images are not necessarily detected. Area-based methods are affected by the intensity distributions of the images. In intensity-based methods different electromagnetic reflectance is present, that's why these methods are not good for multi spectral satellite image registration [2]. On the other hand, the featured-based methods do not depend on the distribution of image intensity values. As an alternative, they use salient features which are extracted from two images, in this scenario it works more suitable where intensity changes and geometrical deformation are encountered. These feature based techniques have been widely used in remote sensing image registration.

In the literature, a number of registration techniques have been proposed, especially for use on satellite images. Yi et al. [2] proposed a SIFT [3] based multi-spectral remote image registration technique which is actually similar to the SURF technique [4]. Song and Zhang [5] presented a method to optimize SURF by defining a similarity measure function based on trajectories generated from Lissajous-figures. They aimed at increasing the feature matching rate. Lee [6] proposed a technique for registering remote sensing images which involved carrying out Haar Wavelet Transform (HWT) before applying the SURF algorithm to the images. Wahed et al. [7] proposed a technique in which median filtering is applied to remote sensing images before performing the SIFT algorithm to register the images. El-Rube et al. [8] presented a technique combining SIFT and multi-scale wavelet transform to register satellite imageries. The control points (or interest points) are selected using three levels of wavelet transform. Manera et al. [9] used the SURF technique to register digital images acquired from digital cameras attached to an unmanned aircraft.

In our previous work [10], a technique to register LandSat-8 satellite images using a combination of well-known image processing algorithms was proposed. In the feature extraction phase, interest (key) points are obtained by means of SIFT and SURF technique. For feature matching, RANSAC algorithm [11] is used. After the application of linear gradient alpha blending method, final image is created. This preciously proposed technique cannot overcome registering more than two mosaic (tile) image because of memory capacity. So, we suggest new methodology to solve this problem in this paper.

Nowadays the procedure based on soft computing techniques such as artificial neural network (ANN), genetic algorithm and fuzzy logic etc. are used for image stitching. Generally Network Architectures are classified into two main classes: First approach is feed-forward networks in which links have no loops (e.g., multilayer perceptron (MLP) and radial basis function neural networks (RBF) .Second approach base on recurrent networks in which loops occur (e.g., self-organizing maps (SOM) and Hopfield networks).Similarly there are different computational techniques used for image stitching for example Radial basis functions [12, 13], self-organizing maps [14], Hopfield networks [15].Correspondingly other proposed technique in this scenario [16] based on a three-layer neural network to determine the registration matrix for 3D surface image stitching. Li and et al. [17] described the use of image regions which lie on an application of pulse-coupled neural network to multi-sensor image fusion problem. Shang and et al. [18] used principal component analysis (PCA) neural network for CT-MR and MR-MR registration. Zhang and et al. [19] illustrate a 3D surface-based rigid registration system for image-guided surgery on bone structures. Sharma and et al. [20] proposed an algorithm that finds the major overlapping area in the images to be mosaicked using neural network (Kohonen's self-organizing Map (KSOFM)). These approaches are mainly applied on medical images rather than satellite images.

3   Architecture

Image registration is the process of aligning two images of a particular area. They correspond to each other on an exact pixel-by-pixel basis [21]. Type of image is an important feature in image registration. Method of image registration varies according to the type of image. This study contains remote sensed images. These types of images usually contain two types of distortion which are radiometric distortion and geometric distortion. In this section, we explain the way of correcting geometric distortion.

In this paper, registration of high-resolution satellite images consists of seven steps. These steps can be listed as; (1) reading image with geographic corner coordinate as latitude, longitude, (2) the calculation of pixel values corresponding to latitude longitude, (3) selection of the most precise value to be calculated in previous step, (4) calculation of width and height of the image to be registered, (5) corner coordinate as (x,y) of each tile image, (6) according to coordinate to be calculated in previous step blending of each tile image to final image using lighten method and finally (7) saving image as jpeg format. This architecture can be seen in Fig. 1 schematically. We will explain these steps in detail later on.

In the first step, high-resolution images to be registered and their geographic corner coordinates as NW, NE, SW and SE are read. These images are obtained from LANDSAT-8 satellite launched by NASA more recently. Each corner coordinate consists of latitude and longitude. For those images, memory is allocated. The more it has been registered number of satellite image, the more it has been allocated amount of memory. More specifically, we can say that if the number of pixel of the satellite image increases, it is necessary to have more memory space. The memory should be used effectively while reading image. To calculate memory space of any image, its dimension and every pixel size are taken into consideration. For example memory space of Fig. 3 is 446 MB (7641 x 7651 x 8 = 467690328 byte). If the size of one image to be registered is about 446 MB, four images are the size of 4x446 = 1784 MB (about 2 GB). Therefore, 2 GB of memory space should be free. Nowadays, it is too hard for a computer to allocate 2 GB memory size for only one application. Unused objects in memory should be de-allocated for optimal usage. This process runs automatically in Java by garbage collector. Although it is not highly recommended, application developers can call the garbage collector. To work actively this process, references of objects to be de-allocated must be removed from memory. If memory has that object reference, the object will continue to occupy memory space. Therefore, firstly the object reference should be removed then garbage collector is called to de-allocate memory space.

In the second step, equivalent of each pixel is calculated as latitude and longitude by using Equation (1) and (2). To register satellite images correctly, their resolutions should be same. In the third step, pixel values belonging to each satellite image are compared with each other and the most precise pixel value is selected to reduce the error rate.

$$X = \text{image\_width} / (\text{NE\_latitute} - \text{SW\_latitute}) \qquad (1)$$
$$Y = \text{image\_height} / (\text{NW\_longitute} - \text{SE\_longitute}) \qquad (2)$$

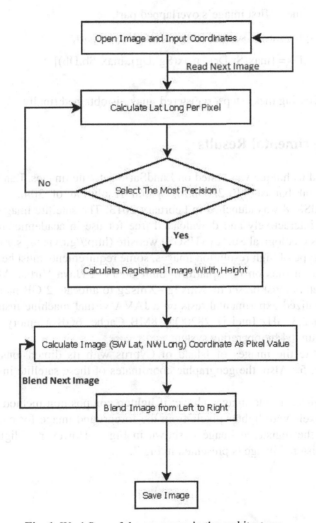

Fig. 1. Workflow of the processes in the architecture

In the fourth step, width and height of registered image are calculated. To calculate image width, it is necessary to determine the most left and the most right longitude. Similarly, to calculate image height, it is necessary to determine the most bottom and the most top latitude. In fifth step, top left point (SW_latitute, NW_longitude) of each satellite tile image is converted into x and y coordinates to find starting point of each tile image on registered entire image. In the sixth step, each satellite tile image is drowned to the registered image by starting from those points. If these tile images are registered directly, a part of black background overlaps with other one part of image which will not be a successful registration. To overcome this problem, color of over-lapped part should be set correctly. While blending overlapped parts of image, max (RGB) function is used as shown (3). This method is called "lighten".

R = Resulting pixel value,

S = Pixel value of first image's overlapped part,

D = Pixel rgb value of second image's overlapped part,

$$R = max(S,D) = [max(Sr,Dr), \ max(Sg,Dg), \ max(Sb,Db)] \tag{3}$$

After processing these steps, registered image is obtained finally.

4   Experimental Results

The proposed technique was tested on LandSat-8 satellite images. LandSat-8 OLI has 8 multispectral bands (440-2200nm), spatial resolution of 30m, and a swath of 185km. LandSat-8 was launched in February 2013. The satellite images obtained can be searched interactively and downloaded free for use in academic studies from the United States Geological Survey (USGS) website (http://glovis.usgs.gov/). In order to register this type of high resolution images, some requirements must be accomplished. This application runs on java platform built on JVM (Java Virtual Machine). Also, JVM parameters must be set to–Xmx2g –Xms2g to allocate 2 GB memory space for JVM. We realized experimental tests on a JAVA virtual machine installed on a ma-chine with a 2.3 GHz Intel i7 2820QM 8MB Cache, 6GB Memory, and Windows Home Premium 64 bit operating system.

The satellite tile images of Island of Cyprus with its dimensions are shown in Fig. 2 to Fig. 5. Also, the geographic coordinates of these satellite images are listed in Table 1.

While registering tile images above; if lighten composition method -replacing tar-get image pixels with lighter pixels from the foreground image for overlapped parts- is not used, the registered image is shown in Fig 6. However, if lighten method is used, the registered image is presented in Fig 7.

Fig. 2. Tile-1, North-West (7641x7651)

Fig. 3. Tile-2, North-East (7651x7801)

Fig. 4. Tile-3, South-West (7641x7791)

Fig. 5. Tile-4, South-East (7631x7431)

Table 1. The geographic coordinates of part satellite images of Island of Cyprus

Images	North West		North East	
	Latitude	Longitude	Latitude	Longitude
Fig. 2	37.0960	32.72594	36.7015	34.80635
Fig. 3	37.0960	34.27671	36.7015	36.35717
Fig. 4	35.6625	32.32869	35.2720	34.37137
Fig. 5	35.6625	33.86668	35.2720	35.90891
	South East		South West	
	Latitude	Longitude	Latitude	Longitude
Fig. 2	34.9760	34.28345	35.3695	32.2481
Fig. 3	34.9760	35.8343	35.3695	33.79888
Fig. 4	33.5447	33.86304	33.9347	31.86247
Fig. 5	33.5448	35.40051	33.9348	33.40038

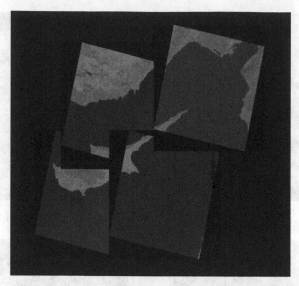

Fig. 6. High-resolution satellite image (13465x13181) before performing Lighten method

Fig. 7. High-resolution satellite image (13465x13181) obtained after performing Lighten method

5   Conclusion

Image registration provided an important element in data processing for remote sensing in the midst of many applications by means of wide range of solutions. Regardless of substantial exploration, the field has not yet settled on a definitive solution. In many applications the numbers of questions still need explanation for appropriative and efficient results. Satellite image mosaicking, a process of stitching or aligning several satellite images to produce a single large scale and high resolution image, is thus an important scheme to extend the usability of satellite imagery in practical applications. However, speediness for image matching, at the same time, assuring the accuracy, i.e. precision, is a key question in the technology of image matching.

This paper proposed an architecture utilizing domain-specific knowledge of geometric transformations and image content. We have used the coordinates of the reference systems, their conversion to screen pixel addresses, etc. Thus, we have increased the performance significantly.

In the future, we will enhance the scalability of the overall system by using physical storages as caches. Thus, a set of high resolution satellite images summing up to terabyte will be able to handle efficiently. We will also consider ANN with proposed technique to obtain more accurate results.

Acknowledgments. This work has been supported by Kocaeli University Scientific Research and Development Support Program (BAP) in Turkey. The project number is 2013/012.

References

1. Brown, L.G.: A survey of image registration techniques. ACM Computing Surveys (CSUR) 24, 325–376 (1992)
2. Yi, Z., Zhiguo, C., Yang, X.: Multi-spectral remote image registration based on SIFT. Electronics Letters 44, 107–108 (2008)
3. Lowe, D.G.: Distinctive image features from scale-invariant keypoints. International Journal of Computer Vision 60, 91–110 (2004)
4. Bay, H., Tuytelaars, T., Van Gool, L.: SURF: Speeded up robust features. In: Leonardis, A., Bischof, H., Pinz, A. (eds.) ECCV 2006, Part I. LNCS, vol. 3951, pp. 404–417. Springer, Heidelberg (2006)
5. Song, Z.L., Zhang, J.: Remote sensing image registration based on retrofitted SURF algorithm and trajectories generated from Lissajous figures. IEEE Geoscience and Remote Sensing Letters 7, 491–495 (2010)
6. Lee, S.R.: A coarse-to-fine approach for remote-sensing image registration based on a local method. International Journal on Smart Sensing and Intelligent Systems 3, 690–702 (2010)
7. Wahed, M., El-tawel, G.S., El-karim, A.G.: Automatic Image Registration Technique of Remote Sensing Images. International Journal of Advanced Computer Science and Applications 4, 177–187 (2013)
8. El-Rube, I., Sharkas, M., Salman, A., Salem, A.: Automatic Selection of Control Points for Remote Sensing Image Registration Based on Multi-Scale SIFT. In: International Conference on Signal, Image Processing and Applications (SIA), vol. 21, pp. 46–50. IACSIT Press, Chennai (2011)
9. Manera, J.F., Rodrigez, L., Delrieux, C., Coppo, R.: Aerial image acquisition and processing for remote sensing. Journal of Computer Science & Technology 10, 97–103 (2010)

10. Sayar, A., Eken, S., Mert, U.: Registering landsat-8 mosaic images: A case study on the Marmara Sea. In: Processing of 10th International Conference on Electronics, Computer and Computation (ICECCO), Ankara, Turkey, pp. 375–377 (2013)
11. Fischler, M.A., Bolles, R.C.: Random sample consensus: a paradigm for model fitting with applications to image analysis and automated cartography. Graphics and Image Processing 24, 381–395 (1981)
12. Davis, M.H., Khotanzad, A., Flaming, D.P.: 3D image matching using radial basis function neural network. In: Processing of WCNN 1996: World Congress on Neural Networks, pp. 1174–1179 (1996)
13. Fornefett, M., Rohr, K., Stiehl, H.S.: Radial basis functions with compact support for elastic registration of medical images. Image and Vision Computing 19, 87–96 (2001)
14. Sabisch, T., Ferguson, A., Bolouri, H.: Automatic registration of complex images using a self organizing neural system. In: Proc. of 1998 Int. Joint Conf. on Neural Networks, pp. 165–170 (1998)
15. Banerjee, S., Majumdar, D.D.: Shape matching in multimodal medical images using point landmarks with Hopfield net. Neurocomputing 30, 103–106 (2000)
16. Liua, H., Yan, J., Zhang, D.: Three-dimensional surface registration: A neural network strategy. Neurocomputing 70, 597–602 (2006)
17. Li, M., Cai, W., Tan, Z.: A region-based multi-sensor image fusion scheme using pulse-coupled neural network. Pattern Recognition Letters 27(16), 1948–1956 (2006)
18. Shang, L., Cheng Lv, J., Yi, Z.: Rigid medical image registration using PCA neural network. Neurocomputing 69, 1717–1722 (2006)
19. Zhang, J., Ge, Y., Ong, S.H., Chui, C.K., Teoh, S.H., Yan, C.H.: Rapid surface registration of 3D volumes using a neural network approach. Image and Vision Computing 26, 201–210 (2007)
20. Sharma, S., Tuli, H., Nagar, S., Dhir, T., Tayal, S.: Using Self-Organizing Neural Network for Image Mosaicing. Advanced Applications of Electrical Engineering, pp. 76–80 (2009)
21. Zagorchev, L., Goshtasby, A.: A Comparative Study of Transformation Functions for Nonrigid Image Registration. IEEE Trans. Image Processing 15(3), 529–538 (2006)

Learning User Models in Multi-criteria Recommender Systems

Marilena Agathokleous and Nicolas Tsapatsoulis

30, Arch. Kyprianos str., CY-3036, Limassol, Cyprus
mi.agathokleous@edu.cut.ac.cy, nicolas.tsapatsoulis@cut.ac.cy

Abstract. Whenever people have to choose seeing or buying an item among many others, they are based on their own ways of evaluating its characteristics (criteria) to understand better which one of the items meets their needs. Based on this argument, in this paper we develop personalized models for each user, according to their ratings on specific criteria, and we use them in multi-criteria recommender systems. We assume the overall ranking, which indicates users' final decision, is closely related to their given value in each criterion separately. We compare user models created using neural networks and linear regression and we show, as expected from the implicit nonlinear combination of criteria, that neural networks based models achieve better performance. In continue we investigate several different approaches of collaborative filtering and matrix factorization to make recommendations. For this purpose we estimate users' similarity by comparing their models. Experimental justification is obtained using the Yahoo! Movie dataset.

Keywords: User modeling, Multi-criteria recommender systems, Collaborative filtering, MCDA, Matrix factorization.

1 Introduction

In order to know if a consumer is going to buy or see a product, it is important to understand first the decision making process, which he/she follows. Decision making can be considered as the cognitive process, which leads to the selection of a course of action between several alternative scenarios. The output of this process includes a final decision, which can be an action or an opinion of choice [11]. Decision making depends on several factors including past experience [9], cognitive biases based on observations [16] as well as age and other demographic differences [4]. Influenced by these factors, each person decodes differently the received stimuli and evaluates situations, objects and interprets services with a unique way, depending on how they perceive their various characteristics. By characteristics, here, we consider individual criteria, which a person usually takes into account in order to conclude in his/her final decision. In item purchasing the most common criteria are price and (measures of) quality. A product is difficult to have the lowest price and the highest measure of quality simultaneously; therefore these criteria are usually in conflict. Furthermore, people realize things

V. Mladenov et al. (Eds.): EANN 2014, CCIS 459, pp. 205–216, 2014.
© Springer International Publishing Switzerland 2014

by their own perspective. Thus, the way they measure high quality and low price is highly subjective. For example, a rich and a poor person perceive the benchmark of price differently; thus, a value considered as low by the rich might be perceived as high by the poor.

Knowing a person's decision making process, gives us the opportunity to understand how they think and evaluate multiple criteria to make the final decision of what item they would like to see or buy. Modeling the evaluation process of each user individually, makes a system able to know better what items fit his/her preferences and allows it to carry out recommendations for products or services that are likely to be absorbed by him/her. A recommendation algorithm that makes use of accurate user models potentially increases its effectiveness since it exploits the knowledge about the basic attributes that attract users to choose an item and recognizes users' taste ensuring a more personalized understanding of them.

In this paper we investigate: (a) whether a neural network based user model can be used to accurate predict the overall evaluation score of an item given the (individual) criteria ratings, (b) to which extent the multivariable function of the overall evaluation score can be linearly approximated using regression models, (c) the effectiveness of matrix factorization techniques on multi-criteria recommender systems that utilize learned user models, (d) whether user model comparison sufficiently captures user preference similarities, (e) to which extent user clustering based on the learned user models improves recommendation accuracy.

For our experiments we used the well-known Yahoo! Movies dataset to allow comparisons with other similar approaches. Both matrix factorization and collaborative filtering methods were examined while the utility matrix used for recommendations was composed either from the original overall item ratings of the actual users or by the predicted by user model ratings.

2 Background and Problem Formulation

The web is deluged with countless information; making navigation very difficult and confused for the most users [13]. That renders Recommender Systems (RSs) very important and necessary. RSs are software tools and techniques, which help users to identify what they really need from a sheer volume of products or services, by providing them personal recommendations. The recommendations concern items or services that could be exploited by a user.

The recommendation problem can be formulated as follows: Let C be the set of users and let S be the set of all possible items that the users can recommend, such as books, movies, restaurants, etc. Let also u be a utility function that measures the usefulness (as may expressed by user ratings) of item s to user c, i.e., $u : C \times S \rightarrow \Re$. The usefulness of all items to all users can be expressed as a matrix U with rows corresponding to users and columns corresponding to items. An entry $u(c, s)$ of this matrix may have either positive value indicating the usefulness (rating) of item s to user c or a zero value indicating that the usefulness $u(c, s)$ has not been evaluated. Although there are several cases where the rating scale is different than the one mentioned above and includes negative values (and as a result the non-evaluated items cannot be represented by the zero value) it is always possible to transform the rating scale in an interval $[l_v \ h_v]$ where both l_v and h_v are greater than zero. The recommendation problem can be seen as the estimation of zero values of matrix U from the non-zero ones. Under this perspective, matrix factorization techniques gained much attention and proved to be highly effective in real-world recommended systems [17,12]. The result of matrix factorization of utility matrix U is, therefore, a matrix $\hat{U}$ whose elements $\hat{u}(c, s) \in [l_v \ h_v]$:

$$\hat{U} = mf(U) \tag{1}$$

where mf denotes a matrix factorization technique.

The main problems in matrix factorization methods are the high dimensionality and sparsity of utility matrix U. The former affects heavily the efficiency of recommendation process (time required to provide a meaningful recommendation) while the latter prevents the use of popular matrix factorization techniques such as Singular Value Decomposition (SVD) which are not effective in sparse matrices [3]. Thus, usage of alternative iterative factorization techniques including Alternative Least Squares (ALS) [12] and Stochastic Gradient Descent (SGD) [17] were proposed. In any case the high dimensionality of utility matrix U and the difficulty of making matrix factorization techniques adaptive (i.e., making a new estimation of utility matrix using the previous one along with the new ratings entering the system) kept classic recommendation techniques alive.

According to Jannach [8] RSs are broadly divided into six main categories : (a) Content-based, (b) Collaborative Filtering, (c) Knowledge based, (d) Community based, (e) Demographic and (f) Hybrid. The most common type is Collaborative Filtering (CF) [10]. CF considers that if two users evaluate the same items in a similar way they are likely to have the same 'taste' and, therefore, RSs can make recommendations between them. The CF approach requires some similarity $sim(c, c')$ between users c and c' to be computed based on the items that both of them evaluated with respect to their usefulness [15]. The most popular approaches for user similarity computation are Pearson correlation and Cosine-based metrics. Both of these methods produce values $sim(c, c') \in [-1 \ 1]$. By computing the similarity of all users in pairs we create the similarity matrix $M \in \Re^{N_C x N_C}$ (N_C is the number of users). Zero values of matrix M may correspond to either zero similarity, or, to users with no commonly evaluated items. The influence of a user can be calculated by taking the sum across the corresponding row or column of matrix M. The higher this sum is the more influential the user is.

Traditional RSs based their recommendations on a single criterion. When they examine an item for its utility to a specific user, they consider as the only criterion item's overall utility score. This score is estimated by the methods mentioned above. In several cases, however, additional information is available. This information can be either of contextual type [2] or of the form of individual ratings of item's characteristics [1]. In both cases additional knowledge about users is gained allowing more personalized recommendations. For instance a travel RS could recommend a holiday package in winter very different from a package in summer, when it is taking into account the temporal context. Also, a holiday package offered to a couple can vary greatly from another, which is addressed to a family. In addition the holiday packages for couples may differ from one couple to the other. Thus, Context-Aware RSs (CARS) try to estimate the utility function $u : C \times S \times Cxt \to \Re$, where Cxt is the set of contexts.

On the other hand, a recommendation problem can be treated as multi-criteria decision making problem, by exploiting the individual ratings of items' characteristics. MCDA is a well-established field of Decision Science that might be structured to support the decision makers in the decision making process, by analysing their options and modeling their value system [5]. Multi-Criteria RSs can use both the ratings of the criteria and the overall utility score to proceed in recommendations. Multi-criteria approaches are well established in marketing research emphasizing on analysis of individual customers' decision making process. According to Hair et al. [6]: "Conjoint analysis is a multivariate technique developed specifically to understand how respondents develop preferences for any type of object (products, services, or ideas). It is based on the simple premise that consumers evaluate the value of an object (real or hypothetical) by combining the separate amounts of value provided by each attribute. Moreover, consumers can best provide their estimates of preference by judging objects formed by combinations of attributes."

In multi-criteria recommender systems the recommendation process is modified to make use of the availability of individual ratings on items' characteristics. In order to formally describe the challenges faced by multi-criteria RSs let us denote with $r_i^{c,s}, i = 1...k$ the individual ratings given by user c to item s to evaluate its k distinct characteristics. The user c also evaluates the overall utility value of item s through the value $u(c,s)$. Obviously individual ratings $r_i^{c,s}$ and overall utility score $u(c,s)$ are strongly related. Thus we can write:

$$u(c,s) = f(r_1^{c,s}, ..., r_k^{c,s}) \tag{2}$$

Since this relation is user dependent we can modify eq. 2 as follows:

$$u(c,s) = f_c(r_1^s, ..., r_k^s) \tag{3}$$

where $r_i^s, i = 1...k$ are the individual ratings of item s and f_c is a multivariable function indicating the personal way user c combines individual ratings of an item into item's overall utility score.

Accurate personalized recommendation benefits from learning function f_c for every user c. This is the basic aim of the current work which suggests a nonlinear approximation of functions f_c using Neural Networks. So far functions f_c were approximated via linear regression models, i.e., f_c is assumed to have the following form:

$$u(c,s) = \sum_{i=1}^{k} w_i^c \cdot r_i^s + w_0^c \tag{4}$$

where $w_i^c, i = 0...k$ are the learned coefficients. Thus, in vector space model representation user c is modeled by the following vector:

$$\boldsymbol{w^c} = [w_0^c \ w_1^c \ ... \ w_k^c]^T \tag{5}$$

In the collaborative filtering approach of multiciteria recommender systems the similarity between users c and c' can be either computed through the traditional way (based on the items that both of them evaluated with respect to their overall usefulness) or by comparing their models, i.e., computing $sim(\boldsymbol{w^c}, \boldsymbol{w^{c'}})$. In the more general case where the users are modeled in a non-linear fashion, as we do in this paper, we suggest computing the similarity of users c and c' by comparing the corresponding user functions, i.e., by calculating $sim(f_c, f_{c'})$.

In matrix factorization approaches it is not clear which is the best way to get advantage of the availability of individual ratings of item characteristics. One way is to extend utility matrix to a 3D tensor T with items $t(c, s, i)$ indicating the evaluation of the i-th characteristic of item s by user c. The problem in this case is that the overall utility evaluation $u(c, s)$ is totally ignored. Furthermore, factorization of sparse tensors is even more complex and computationally intensive than matrix factorization.

In this paper we investigate the application of matrix factorization on $\tilde{U}$. $\tilde{U}$ is computed by replacing the non-zero values of the actual utility matrix U with the ones estimated by the learned user models:

$$[\tilde{U}|\tilde{u}(c, s) = f_c(r_1^s, r_2^s, , ..., r_k^s), u(c, s) \neq 0] \tag{6}$$

Thus, instead of using eq. 1 for solving the recommendation problem we use $\hat{U} = mf(\tilde{U})$.

3   Research Questions and Methodology

In this study we try to advance the research in multiciteria recommender systems by investigating the following research questions: (1) Is non-linear modeling of user functions f_c more effective that linear models $\boldsymbol{w^c}$? (2) In non-linear modeling using Neural Networks how the number of hidden neurons H_N affects user modeling? (3) Can we estimate user similarity by comparing user functions f_c (instead of the traditional approach which involves similarity computation based on commonly evaluated items)? (4) In matrix factorization can we factorize the estimated through the learned models utility matrix $\tilde{U}$ instead of the original utility matrix U?

In the first research question we assume that the overall utility score, which a user gives to an item, is closely related to his/her given value in each individual criterion separately. While for some users the criteria may contribute equally to the overall score, in the majority of cases users give different significance to each individual criterion and combine the criteria in, an unclear even to them, way to produce the final utility score. The question is whether the introduction of non-linearity, in individual criteria combination, leads to better user modeling.

A similar study was made by Jannach et al. [7], who used a tourism platform that provides the ratings of multiple criteria about the accommodation. Regression models that constitute specific aggregation functions for each user and each item were compared. They concluded that regression models learned with a classifier using a support vector machine, outperform linear least squares regression models.

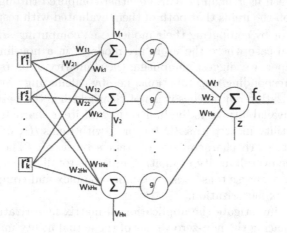

Fig. 1. The model of user function f_c

In the second research question we try to estimate the degree of non-linearity of user function f_c. We expect that f_c is approximately linear, thus, a small numbers of hidden neurons suffice for its modelling. In this perspective fc is given by (see also Figure 1):

$$f_c(r_1^s, r_2^s, ..., r_k^s) = \sum_{j=1}^{H_N} \left\{ w_j \cdot g\left(\sum_{i=1}^{k} w_{ij} r_i^s + v_j \right) \right\} + z \qquad (7)$$

where $g(\cdot)$ is a nonlinear function (usually sigmoid) and w_{ij}, v_j, w_j, z, the modeling parameters learned through the Neural Network and H_N is the number of hidden neurons. Since the number of modelling parameters N_p is heavily dependent on the number of hidden neurons ($N_p = (k + 2)H_N + 1$) it is highly desirable to minimize it.

Traditionally, in collaborative filtering approaches of RSs similarity between users (in order to identify users with similar preferences) is computed by comparing their evaluations on commonly assessed items [10]. Extending this approach to multi-criteria RSs is not straightforward. Again, one should either ignore the overall rating and compare the individual ratings of the commonly evaluated items which results in comparing matrices with dimensions $N_S \times k$ (N_S is the number of items in the RS and k is the number of item characteristics), or ignore the ratings on the individual characteristics and only use the overall rating.

This is the main reason researchers working with multi-criteria RS end up comparing user models in order to estimate similarity of user preferences. In our third research question we investigate whether we can use $sim(f_c, f_{c'})$ for this purpose. We consider that the nonlinear function f_c can be represented through its learned parameters:

$$f^c = [w_{11}\ w_{21}\ \ldots\ w_{k1}\ v_1\ w_{12}\ w_{22}\ \ldots\ w_{k2}\ v_2\ \ldots\ w_{1H_N}\ w_{2H_N}\ \ldots\ w_{kH_N}\ v_{H_N} \tag{8}$$
$$w_1\ w_2\ \ldots\ w_{H_N}\ z]^T$$

$sim(f_c, f_{c'})$ is computed, then, through the following equation:

$$sim(f_c, f_{c'}) = 1 - \frac{||f_c - f_{c'}||}{||f_c|| + ||f_{c'}||} \tag{9}$$

where $||f_c||$ denotes the norm of vector f_c.

Finally, in the fourth research question we investigate whether we can use factorization of utility matrix $\tilde{U}$ (estimated through the learned models) instead of the original utility matrix U and whether this can improve the performance of matrix factorization approaches.

In this work we use feed-forward neural networks, since they are good at fitting functions to the available data. They composed of a set of inputs $r = [r_1, r_2, \ldots, r_k]$, a hidden layer h, consisting of H_N hidden neurons and one output r_o, where r_o is the total rating of each user and the inputs are the values of the k criteria. Each input r_i is weighted by a weight w_{ij}, directed to the $j - th$ hidden neuron and summed (along with a shifting bias v_j) producing a quality, say, y_j. Then a sigmoid function $g(y_j)$ is activated mapping from y_j to h_j, which is the output of hidden neuron j. The outputs of all hidden neurons are weighted by the parameters w_j and summed (along with a shifting bias z) to produce the final output r_o. Training refers to the estimation of weights and biases [14] for every particular user based on his/her individual and overall ratings on already evaluated items.

For comparison we also built linear user models through linear regression as indicated in eq. 10. That is, for a user c the total rating r_o of a movie can be calculated by the given value to each criterion separately as follows:

$$r_o^c = \sum_{i=1}^{k} w_i^c r_i + \varepsilon \tag{10}$$

where w_i^c is the weight of the criterion value r_i for user c, k is the number of the criteria and ε is the intercept of the linear predictor.

4   Experimental Results and Discussion

4.1   Dataset

In this work we used the Yahoo!Movies data (movies.yahoo.com) for experimental investigation of the various research questions mentioned in the previous section. In this dataset users provide preference information on movies based on four different criteria: acting, story, direction and visuals. They also apply a total rating, which summarizes their total evaluation of movie. All criteria ratings were measured in a 13-fold qualitative scale, with $A+$ referring to the highest grade and F representing the worst evaluation grade. For processing purposes, the letters of evaluation were replaced with numbers, so as 1 corresponded to the worst value F and 13 to the best value, $A+$. To be sure that the information for each user is enough, to train a neural network, we took into account only the users who graded more than thirty movies. The resulting experimental dataset includes 239 different users and 976 movies. Even so, the sparsity of data is huge (94.3%), since there are 976 different movies, which are not all rated by all users. The main characteristics of this dataset are summarized in Table 1.

Table 1. Dataset Characteristics of Yahoo!Movies data

#users	#movies	#ratings	#ratings per movie (average)	sparsity
239	976	13286	14	0.9430

4.2 Non-linear Modeling of User Function f_c and the Number of Hidden Neurons

In this experiment we investigated whether a neural network based user model could predict the total rating of a movie, given the criteria ratings. We trained neural networks by randomly dividing the dataset to training set (50% of the movies each user evaluated), validation set (20% of movies each user evaluated) and testing set (30% of movies each user evaluated). As inputs we consider the values of the k criteria of each movie, which rated by each user and as output the total rating given by that user. We started by using one hidden neuron and gradually increased to four hidden neurons (equal to the number of criteria). We compared user modeling via neural networks with the corresponding (to the same user) least squares regression model. For the latter we used the same training and testing sets we used in neural networks. Neural network models were retrained ten times for each user and we kept the average performance of each of them.

Table 2 shows the results of user modelling through Neural Networks and Linear Regression. By winning model we mean the model which better fits a particular user, i.e., estimates better the overall rating based on the individual ones. For instance when we used four hidden neurons in 51,74% of the users Neural Network models predicted better the overall rating while in the remaining 48,26% Linear Regression models performed better.

With respect to the number of winning models it appears that the advantage of using neural networks (non-linear modelling) instead of simple linear regression models is not high. This indicates that the majority of users combine

Table 2. The results of modeling users w.r.t winning models as a function of the number of hidden neurons (HN). The number of hidden neurons refers only to the neural network.

Modeling Method	# Hidden Neurons			
	1	2	3	4
Neural Network	0.5071	0.5155	0.5159	0.5174
Linear Regression	0.4929	0.4835	0.4841	0.4826

the individual ratings through a simple linear weighting scheme to conclude to the overall rating. In an effort to investigate this further we compared neural network and linear regression models in terms of recommendation effectiveness. Table 3 shows this comparison in terms of Mean Absolute Error (MAE). In our experiments a variation of MAE [10,13] is computed with the aid of Frobenius norm as indicated in eq. 11.

$$MAE = \frac{1}{N_c} \sum_{i=1}^{N_c} |U - \hat{U}| \tag{11}$$

where $\hat{U}$ is the estimation of utility matrix U while N_c is the total number of users. Keep in mind that U includes only the overall ratings and not the ratings of the individual criteria.

Table 3. Evaluation of user modeling process in terms of MAE as a function of the number of hidden neurons (HN). The number of hidden neurons refers only to the neural network, thus, the performance of linear regression independent of this number.

Modeling Method	# Hidden Neurons			
	1	2	3	4
Neural Network	0.6511	0.6470	0.6449	0.6400
Linear Regression		0.6639		

We observe, again, that the difference in recommendation effectiveness, in terms of MAE, between the two modeling approaches is too small. Thus, we confirm that user functions f_c can effectively approximated via linear models and the gain when using non-linear approximation is low.

Concerning the hidden neurons (note that the number of hidden neurons refers only to the neural network and does not affect the performance of linear regression) we observe, in both Tables 2 and 3, that by increasing their number the performance of neural network based modelling improves slightly. This is in accordance with our previous conclusion that user function f_c is approximately linear.

4.3 Collaborative Filtering and Matrix Factorization Using the User Function f_c

In the next two experiments we used the non-linear modeling of user function f_c in the recommendation process. We examined several different techniques falling under two main categories: collaborative filtering (see [15] for more details) and matrix factorization (see also [3]). The aforementioned methods were applied on two utility matrices. The first one contains the predicted total ratings of all users (found with the aid of user models f_c), while the second contains the original total ratings that users gave. We show below that the performance is almost the same despite the utility matrix used. This indicates, again, that user modelling is effectively tackled through the functions f_c and the predicted overall ratings are only slightly different from the original values.

In order to measure the overall recommendation effectiveness we use the following accuracy metric:

$$A = \frac{||U - \hat{U}||}{||U|| + ||\hat{U}||} \tag{12}$$

where $||U||$ denotes the Frobenius norm of matrix U.

Tables 4 and 5 show the performance, in terms of the accuracy metric defined above, for various Collaborative Filtering (CF) and Matrix Factorization (MF) methods respectively. While our dataset is too sparse and matrix factorization techniques perform better in such datasets, the clustering based methods (k-Means, k-NN, Fixed Distance) show better performance than ALS and SVD. This observation indicates that comparing user functions f_c to assess the similarity between users is quite effective since clustering of users was based on this similarity. However, the overall best score is achieved by SGD showing that matrix factorization properly adapted to very spare matrices outperforms CF methods. One the other hand traditional MF methods, like SVD, are not effective in sparse data.

Table 4. Comparison of various CF techniques when the similarity between users is estimated using the traditional approach $(sim(c, c'))$ and by comparing learned models $(sim(f_c, f_{c'}))$. Shown values refer to accuracy A as indicated in eq. 12.

	Method			
	Nearest Neighbor	k-Means	K-NN	Fixed Distance
$sim(c, c')$	0.8921	0.1606	0.1792	0.1864
$sim(f_c, f_{c'})$	0.8930	0.1610	0.1796	0.1860

5 Conclusion

In this article we dealt with the problem of recommendation in multi-criteria recommender systems (MC-RS), trying to investigate parameters which would increase their effectiveness. We proposed modeling the way users combine individual criteria to arrive to an overall rating, for an item, through neural networks

Table 5. Comparison of various matrix factorization techniques applied on the original utility matrix U and on the estimated through learned models utility matrix $\tilde{U}$. Shown values refer to accuracy A as indicated in eq. 12.

	Method		
	SGD	ALS	SVD
Factorization of U	0.1244	0.6019	0.6035
Factorization of $\tilde{U}$	0.1236	0.6015	0.6034

assuming that this decision making process is more complex than it appears to be in the corresponding literature. We then investigated, experimentally, four research questions related with this modeling.

In the first research question we assumed that users' decision making process is nonlinear, thus, neural network models would perform better than linear regression. Experimental results showed that the difference in performance between the two methods is negligible, thus, linear modeling, which is less complex, is sufficient for user modeling in MC-RS. In the second experiment we examined whether the number of hidden neurons affects user modeling and recommendation effectiveness. In accordance to the previous conclusion it appears that user models are not so complex to require more than a single hidden neuron. In the last two research questions we investigated (i) whether user similarity can be estimated by comparing user models (instead of the traditional approach which estimates user similarity based on the commonly evaluated items), and (ii) whether in matrix factorization techniques the predicted utility matrix can be used instead of the original one. Experimentation showed that in both cases the answer is positive, showing that user modeling (despite the method adopted) is a valuable tool for increasing the effectiveness in MC-RSs.

We also tested several recommendation techniques to assess their performance in MC-RSs. We found that the Stochastic Gradient Descent (SGD) method outperforms every other technique remaining insensitive to the sparseness of the dataset. On the other hand, the clustering based methods proved to be surprisingly robust and in most cases outperform their matrix factorization counterparts.

In the near future we plan to test the proposed method in more datasets especially to ones having more criteria to confirm that the conclusions drawn here are still valid. In addition, we will test the effectiveness of user modeling in non-sparse datasets, such as in vote recommendation (see [3]).

References

1. Adomavicius, G., Kwon, Y.: New Recommendation Techniques for Multicriteria Rating Systems. IEEE Intelligent Systems, 48–55 (2007)
2. Adomavicius, G., Tuzhilin, A.: Context-Aware Recommender Systems. In: Ricci, F., Rokach, L., Shapira, B., Kantor, P.B. (eds.) Recommender Systems Handbook, pp. 217–250. Springer US (2011)

3. Agathokleous, M., Tsapatsoulis, N.: Voting Advice Applications: Missing Value Estimation Using Matrix Factorization and Collaborative Filtering. In: Papadopoulos, H., Andreou, A.S., Iliadis, L., Maglogiannis, I. (eds.) AIAI 2013. IFIP AICT, vol. 412, pp. 20–29. Springer, Heidelberg (2013)
4. Bruine de Bruin, W., Parker, A., Fischhoff, B.: Individual Differences in Adult Decision-Making Competence (A-DMC). Journal of Personality and Social Psychology 92, 938–956 (2007)
5. Dodgson, J.S., Spackman, M., Pearman, A., Phillips, L.D.: Multi-criteria analysis: a manual. Department for Communities and Local Government: London (2009) ISBN 9781409810230
6. Hair, J.F., Black, W.C., Babin, B.J., Anderson, R.E.: Multivariate Data Analysis, 7th edn. Prentice Hall (2010)
7. Jannach, D., Gedikli, F., Karakaya, Z., Juwig, O.: Recommending hotels based on multi-dimensional customer ratings. In: International Conference on Information and Communication Technologies in Tourism, pp. 320–331. Springer (2012)
8. Jannach, D., Zanker, M., Felfernig, A., Friedrich, G.: Recommender Systems: An Introduction. Cambridge University Press (2010)
9. Jullisson, E.A., Karlsson, N., Garling, T.: Weighing the past and the future in decision making. European Journal of Cognitive Psychology 17(4), 561–575 (2005), doi:10.1080/09541440440000159.
10. Herlocker, J.L., Konstan, J.A., Riedl, J.T.: An empirical Analysis of Design Choices in Neighborhood-Based Collaborative Filtering Algorithms. Information Retrieval 5(4), 287–310 (2002)
11. Reason, J.: Human error. Cambridge University Press, New York (1990)
12. Salakhutdinov, R., Mnih, A.: Probabilistic Matrix Factorization. In: Advances in Neural Information Processing Systems (NIPS 2007), pp. 1257–1264. ACM Press (2008)
13. Shardanand, U., Maes, P.: Social information filtering: Algorithms for automating Word of mouth. In: ACM CHI 1995 Conference on Human Factors in Computing Systems, pp. 210–217. ACM Press (1995)
14. Taner, M.T.: Neural networks and computation of neural network weights and biases by the generalized delta rule and back-propagation of errors (1995)
15. Tsapatsoulis, N., Georgiou, O.: Investigating the Scalability of Algorithms, the Role of Similarity Metric and the List of Suggested Items Construction Scheme in Recommender Systems. International Journal on Artificial Intelligence Tools 21(4), 19–26 (2012)
16. West, R.F., Meserve, R.J., Stanovich, K.E.: Cognitive Sophistication Does Not Attenuate the Bias Blind Spot. Journal of Personality and Social Psychology (2012), doi:10.1037/a0028857 (Advance online publication)
17. Zhou, T., Shan, H., Banerjee, A., Sapiro, G.: Kernelized Probabilistic Matrix Factorization: Exploiting Graphs and Side Information. In: SIAM International Conference on Data Mining, pp. 403–414. SIAM / Omnipress (2012)

Fault Classification System for Computer Networks Using Fuzzy Probabilistic Neural Network Classifier (FPNNC)

Karwan Qader and Mo Adda

University of Portsmouth, School of Computing,
Buckingham Building, Lion Terrace, PO1 3HE Portsmouth, Great Britain
{karwan.qader,mo.adda}@port.ac.uk

Abstract. Over the last decade, the world has witnessed the rapid development of networking applications of different kinds, and network domains have become more and more advanced regarding with their level of heterogeneity, complexity and the size. Some obstacles such as availability, flexibility and insufficient scalability have affected the existing centralized network management systems, as networks become more distributed. In this work a Fuzzy Probabilistic Neural Network Classifier (FPNNC) is proposed, comprising a hybrid fault classification algorithm based on Fuzzy Cluster Mean (FCM) with Probabilistic Neural Network (PNN) to classify the detected fault datasets. These results will assist network administrators with a highly effective tool to classify faults that occur in computer network systems, enabling them to take well-informed decisions pertaining to security, faults and performance.

Keywords: Clustering, classification, network faults, fault diagnosis, FCM, PNN, FPNNC.

1 Introduction

With the rapid development of computer network technology, the scale and function of networks is constantly increasing. The increasing importance and complexity of networks led to the development of network fault management as a distinct field, providing support for network administrators with quality services and ensuring that networks work appropriately. Fault diagnosis is a central aspect of network fault management. Since faults are unavoidable in communication systems, their quick detection and isolation is essential for the robustness, reliability and accessibility of the system. In large and complex communication networks, automating fault diagnosis is critical. Because of many factors, including the volume of network information, it is hard to solve network fault problems with traditional tools, rendering intelligent diagnosis a critical method in the process of network fault diagnosis [1].

In the process of network fault diagnosis, both cluster and classifier techniques play a significant role by identifying types and locations of the faults. The use of

V. Mladenov et al. (Eds.): EANN 2014, CCIS 459, pp. 217–226, 2014.

clustering in grouping objects is one of the most commonly used data mining techniques. The resultant groups of objects can help a network administrator to take accurate decisions to protect data communications over a network. The method based on back propagation (BP) technique is most extensively used in intelligent diagnosis method of artificial neural network [2]. Statistics show that 80% of neural network models have adopted BP network or its variants. However, the neuron numbers of the BPs imply layer are attained by experience, not from precise computing of theory, and the BP neural network has several shortcomings, such as falling into local least point easily and needing a long time for training [3].

Network PNN is an extensively used artificial neural network. Its structure is simple and its training succinct. The advantage of PNN lies in finishing the work with the linear study algorithm, which was previously, achieved using nonlinear study algorithm.

This article is organized as follows. The following section reviews related studies, followed by explanation of the proposed algorithm Fuzzy Probabilistic Neural Network Classification (FPNNC), with discussion of the network characteristic parameters and patterns. Section 4 describes the case studies and the source of the datasets. The results are presented and discussed in section 5. Finally, the conclusion of the paper is presented in section 6.

2   Related Work

In recent years, much research has been undertaken to explore network faults, particularly fault diagnosis and management. However, a recent review study has shown that although the trends of fault management and diagnosis have been increasingly explored in recent research papers, most of them do not include contributions about the fault diagnosis in computer network system. This paper addresses this gap by extending prior work focused on faults classification to practical application for computer network system.

A number of studies have been carried out and methods have been proposed in the field of fault detection and classification for semiconductor manufacturing equipment [4]. One such method put forward by [5], detects faulty processes of the semiconductor manufacturing equipment using its data, recognizes anomalies and classifies the root cause of the faults by reading the production equipment data, which consists of all such information; albeit this can be quite useful, it is challenging to study due to the complexity and volume of data [6]. Hence, this paper explores Modular Neural Network modelling, whereby data from production equipment is aggregated into associated subsystems, enabling Fault Detection and Classification(FDC) using Dempster-Shafer (DS) method to consider the ambiguities in fault detection. The method employs Radio Frequency (RF) power source module probing, which is advantageous for detecting the chamber leak simulation and helps in classifying the faults by evaluating the RF probe voltage signals. This paper was successful in justifying the use of D-S theory by successful fault detection at subsystem level, with no missed alarms [7] [8].

Fault detection is one of the most important network management tasks, as analysed by [9], to propose a statistical method based on Wiener filter to capture the abnormal changes in the behaviour of the MIB variables. The algorithm of the study took data from two different scenarios and four different case studies. Such an analysis provided the manager node of the network high level of information instead of huge data volume [9]. The study in [10] of Sensor Fusion and Sensor Fault Detection with Fuzzy Clustering presented an effective approach for multi-sensor fusion and fault diagnosis, which makes use of FCM algorithms for separating the signals. The fusion engine in turn generates a fused signal based on the concept of centre of gravity (COG) de-fuzzification method. In this approach, the sensor fault detector is designed from the total fused signal residual and the output of the sensors. The results of the simulation showed a clear improvement in the fault detection accuracy [10]. A two-phase approach for measuring the performance of the cluster based internet services was presented by [11]. The first phase of the methodology employs the fault-injection approach for measuring the impacts of faults on the network performance while the second phase makes use of analytical models to assess the network performance by combining the measurements of first phase and the fault loads [12]. Such a two-phased approach lets the evaluator study how the servers respond to various design-related decisions, rate of faults and other factors. Four versions of PRESS web servers were tested against five fault classes to measure the performance of the servers in different scenarios [11] [13].

The area of smart networks has been the subject of good amount of research and review recently because of the concept of computational intelligence, which is incorporated into smart networks [14]. Because of computational intelligence, smart networks are capable of detecting their faults and classifying them [15]. Keeping in view this capability of smart networks, [16] presented two techniques for fault detection and classification in power transmission lines in smart networks. The techniques proposed are based on Quarter Sphere Support Vector Machine (QSSVM). The first approach makes use of Temporal Attribute QSSVM, utilizing the temporal, and attributes correlations of the measurements of the transmission lines for detection of faults during the stage of transient. The second approach makes use of Attribute QSSVM (A-QSSVM) and takes into account attribute correlations for the automatic fault detections and classifications [17], the results of these two approaches displayed accuracy in fault detection and classification (as high as 99%), which amounted to a significant reduction in terms of computational complexity compared to traditional techniques making use of multi class SVM for fault detection and classification [18]. Additionally, as compared to the traditional methods, these techniques are still quite unsupervised and can be available for implementation on the existing fault-monitoring infrastructure for limited online supervision in power systems of smart networks [16].

An ensemble Fault Diagnosis Based on Fuzzy C-means Algorithm of the Optimal Number of Clusters and Probabilistic Neural Network (FCM-ONC-PNN) represented by [19] portrayed the significance of fault diagnosis as a process being followed for maintaining quality of the products in industrial systems and ensuring various aspects like reliability, safety and efficiency from the point of view of operations in many plants.

Finally Fuzzy C-Means Clustering and Feed Forward Neural Network was used by [20] to find the fault-proneness of a software module, focusing on the benefits of the early detection of fault prone software components and how hybrid approach based on fuzzy C-means clustering based approach and feed-forward neural network based approach can be used to find faults. The proposed method of the current study rely on FCM and PNN. The main reasons behind selecting these two techniques refers to their robust factors. Fuzzy Cluster Means (FCM) is the most suitable algorithm among clustering techniques due to its robust characteristics to deal with network fault diagnosis problems, such as handling unclear boundaries of clusters, overcoming high dimensionality problems and its flexibility, especially for traffic analysis, which is an iterative optimal algorithm [21][22]. Likewise, PNN maintains excellent characteristics such as high precision of the non-linear algorithm. The corresponding weights values of PNN are the distribution of the model sample, and also the network does not need training, therefore it can meet the requirements of real-time processing [2]. Moreover, its robust characteristic to sample noises, fast speed of its training data and the accuracy rate of the classification, enhance achieved results to have a higher quality [23].

3   Proposed Network Faults Classification Algorithm (FPNNC)

The Fuzzy Neural Probabilistic Neural Network Classification (FPNNC) is proposed to classify the faults that occur in the computer network system. The developed algorithm comprises the combination of two techniques, Fuzzy Clustering Means (FCM) as unsupervised learning technique and Probabilistic Neural Network (PNN), as shown in fig. 1.

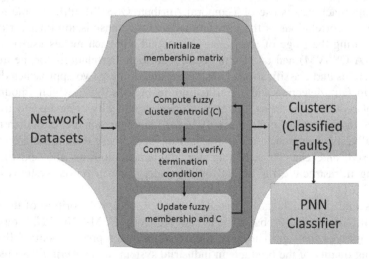

Fig. 1. Schematic overview of the proposed approach (FPNNC)

FCM consists mainly of four phases used to generate clustered feature vectors from the data points of each specific fault; the featured vectors are then forwarded to PNN in order to classify the fault types.

3.1 FCM

Fuzzy C-means (FCM) is one of the most common unsupervised clustering methods that was originally proposed by Bezdek in 1981. As with other clustering techniques, FCM primarily relies on measuring the distance between data points and it uses Euclidean distance to measure similarity between objects. The distance measure helps the algorithm to make decisions to create groups for each data point depending on the similarity and dissimilarity between points. Similar data points are kept in a group known as a cluster. Similarity is assayed with values between 0.0 and 1.0; the value 0.0 indicates highly dissimilar, while 1.0 indicates the highest similarity between objects. As illustrated in figure 1, the system takes the data points as input then generate the clusters based on four main phases.

The Euclidean Distance Function, which is known as objective function J_m is used in FCM to get fuzzy C partition $A = \{A_1, A_2, A_3, ..., A_n\}$ for given dataset $X = \{X_1, X_2, X_3, ... X_n\}$ and number of clusters denoted by "c". The main objective in FCM is to minimize J_m depicted in the following equation in order to get the optimal clusters. As it is an iterative method, it has to achieve better minimization in each iteration.

$$J_m(\mu, V : X) = \sum_{i=1}^{c} \sum_{j=1}^{n} (\mu_{ij})^m \parallel X_j - V_i \parallel^2 \tag{1}$$

Where μ_{ij} represents the membership function of data point X_j in the ith cluster. m is the fuzzifier which acts as the fuzziness controller value; generally m is any real number that is greater 1. V_i is the centre of ith cluster, whereas ||.|| denotes the Euclidean distance.

The equation (1) is used to compute the value of J_m (U, V) and determine the criterion function based on a threshold. Firstly, the membership function and cluster centroids are randomly initialised with some constraints consideration:

$$\sum_{i=1}^{c} \mu_{ij} = 1 \tag{2}$$

In order to optimise the results in each iteration through an iterative process, the fuzzy membership and fuzzy cluster centroids are updated using the equations shown below:

$$V_i = (\sum_{j=1}^{n} (\mu ij)^m X_j) / (\mu_{ij})^m \tag{3}$$

$$\mu_{ij} = [\sum_{k=1}^{c} (\parallel X_j - V_i \parallel^2 / \parallel X_j - V_k \parallel^2)^{1/(m-1)}]^{-1} \tag{4}$$

3.2   PNN

Probabilistic Neural Network (PNN), which was introduced by D.F. Specht in the early 1990s, is one of the common feed-forward neural networks used in classification problems [24]. The general architecture of PNN is organized based on multi-layered feed-forward network into four different layers (input layer, hidden layer, pattern layer or summation layer and output layer), as shown in fig. 2.

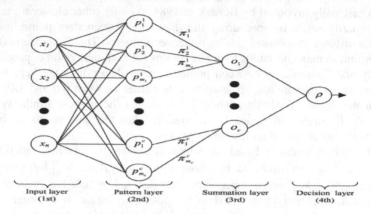

Fig. 2.   Diagram of PNN [25]

The hidden and summation layers, known as the radial basis layer, refers to all equations applied to the input neurons, which calculate the distance from input vectors to training input vectors. Competitive (summation) layer is considered to be another main layer that compares the weighted vectors and select the maximum one to predict the target category. The number of neurons in summation layer is equal to the number of train set neuron.

The Euclidean Distance function used to calculate the distance and to indicate how close the input is to a training sets in the pattern layer as shown below:

$$D_j = \left\lVert W_{ij} - X \right\rVert . b \tag{5}$$

Where b represent to Radial Basis Function (RBF), which can be used to adjust the susceptibility of the radial basis neurons by having different values and set to:

$$b = \frac{\left[-\log(0.5) \right]^{1/2}}{Spread} \tag{6}$$

Spread is the extended coefficient of RBF. Moreover, the Radial Basis Function (RBF) as shown in equation (7), is applied to calculate the probability of each inputted neuron (X) by comparing the input vectors with weight neuron vectors W_{ij}.

$$R_j = radbas \left(D_j \right) \tag{7}$$

Out of several candidate functions of RBF, a Gaussian function has been selected for the proposed work.

$$R_j = e^{-D_j^2} \tag{8}$$

Thus, the output relies on the distance between W_{ij} and X; it inverts with the distance, thus when the distance decreases it increases to reach its maximum while $W_{ij} = X$.

4   The Characteristic of Network Faults

SNMP protocols use Management Information Base (MIB) to obtain knowledge about the status information of computers or devices in network in order to manage objects in the right manner. MIB includes any statistical and status information about each node in the system. It has information about counters at each node interface and uses them to indicate the number of packets or octets that have been sent, dropped, received and delivered. Each device has one or more interface.

The total functions of SNMP rely on MIB, which follows the ISO standards and is defined by RFC 1155, which locates the network devices. In 10 groups of MIB, 10 variables out of 178 MIB variables were selected from interface and IP groups. The variables in these two groups are more likely to have more sensitivity with traffic. Table 1 shows all selected parameters in IF and IP group that enable the system to diagnose faults easily.

Table 1. Characteristic Parameters of Network

No.	Network Parameter
P1	ifInNUcastPkts22
P2	ifInOctets22
P3	ifInUcastPkts22
P4	ifOutNUcastPkts22
P5	ifOutOctets22
P6	ifOutUcastPkts22
P7	ipForwDatagrams22
P8	ipInDeliver22
P9	ipInReceive22
P10	ipOutRequests22

5   Case Study and Results Discussion

The proposed work complements and extends past research work detecting, classifying and collecting network faults. The datasets were captured in two different (heavy and light) scenarios using MIB variables from IP and IF groups for four different types of traffic. The main aims to collect data in two different scenarios is to check the capability of the system in terms of whether it can detect the traffic faults in different environment or not. The two scenarios were created by changing the size of packets and bandwidth. ifInUcastPkts, ifInNUcastPkts, ifOutUcastPkts and ifOutNUcastPkts are more affected by traffics than the other selected variables. In addition, the experiment tested the router as the first gate to attack the server as the main target. The router differs from the server in that its variables are limited to network layer group (IP group). In contrast, in the case of a server, one can read most of the MIB variables, starting from a low-level group such as the interface group, to an application layer group.

The current model is applied on four different types of network traffic, which includes server crash or server link failure, broadcast storm, babbling node, and normal traffic. Matlab was used to give different outputs and results based on FCM and PNN for the available datasets. After all data points were inputted to the system, FCM generated clusters of featured vectors for each dataset. The empirical results show all available clusters when there is a failure in the link to server or server crash.

Consequently, PNN creates a model to classify these vectors properly, which they attain by FCM. First, all target classes' indices are converted to vectors, in order to compare them with the input vectors. After consecutive mathematical calculations, the system starts testing the network on the design input vectors. This make the network simulation and converts its vector outputs to indices.

The proposed system was tested and applied on another three different types of network traffic datasets. The results obtained from the preliminary simulation of the proposed work shows different output patterns for each specific faults.

Overall, these results indicate that the proposed algorithm FPNNC is able to classify all inputted fault datasets properly and provide distinct output patterns for each specific fault traffic. Consequently, it gives a clear vision to network administrator to realize about the location and the type of faults.

6   Conclusion

The key aspect of network fault management is the process of a fault classification, by which it concludes the details of a failure from a set of tested failure indications. This paper has given an account of and the reasons for the widespread use of FCM and PNN in classification problems. The main purpose of the current study was to use new proposed technique (FPNNC) to classify the faults properly in a computer network system. The experiment results show that FPNNC is able to classify the faults in the real-time system and categorize them into different patterns outputs based on the types of fault that occur. Consequently, these results provide a level of security and performance for the network system, helping network administrators to have a clear vision for the problems based on visualization of classified faults and to take decisions faster related to security, performance and faults.

References

1. Fenton, W.G., McGinnity, T.M., Maguire, L.P.: Fault diagnosis of electronic systems using intelligent techniques: a review. IEEE Transactions on Systems, Man, and Cybernetics, Part C: Applications and Reviews 31(3), 269–281 (2001)

2. Gao, Y., Zhou, X.: The design of network fault diagnosis system based on PNN. Paper Presented in 2nd International Conference at the Future Computer and Communication (ICFCC) (2010)
3. Zheng, Q., Qian, Y., Yao, M.: A network event correlation algorithm based on fault filtration. In: Yang, Q., Webb, G. (eds.) PRICAI 2006. LNCS (LNAI), vol. 4099, pp. 864–869. Springer, Heidelberg (2006)
4. Barakat, M., Druaux, F., Lefebvre, D., Khalil, M., Mustapha, O.: Self-adaptive growing neural network classifier for faults detection and diagnosis. Neurocomputing 74(18), 3865–3876 (2011)
5. Hong, S.J., Lim, W.Y., Cheong, T., May, G.S.: Fault Detection and Classification in Plasma Etch Equipment for Semiconductor Manufacturing-Diagnostics. IEEE Transactions on Semiconductor Manufacturing 25(1), 83–93 (2012)
6. Rengaswamy, R., Venkatasubramanian, V.: A fast training neural network and its updation for incipient fault detection and diagnosis. Computers & Chemical Engineering 24(2), 431–437 (2000)
7. Chao, C.S., Liu, A.C.: An alarm management framework for automated network fault identification. Computer Communications 27(13), 1341–1353 (2004)
8. Hong, S.J., Lim, W.Y., Cheong, T., May, G.S.: Fault Detection and Classification in Plasma Etch Equipment for Semiconductor Manufacturing-Diagnostics. IEEE Transactions on Semiconductor Manufacturing 25(1), 83–93 (2012)
9. Al-Kasassbeh, M., Adda, M.: Network fault detection with Wiener filter-based agent. Journal of Network and Computer Applications 32(4), 824–833 (2009)
10. ElMadbouly, E., Abdalla, A., ElBanby, G.M.: Sensor fusion and sensor fault detection with fuzzy clustering. In: International Conference Presented at the Computer Engineering and Systems (ICCES) (2010)
11. Nagaraja, K., Li, X., Bianchini, R., Martin, R.P., Nguyen, T.D.: Using Fault Injection and Modeling to Evaluate the Performability of Cluster-Based Services. Paper presented at the USENIX Symposium on Internet Technologies and Systems (2003)
12. Song, Y.-H., Johns, A., Xuan, Q., Liu, J.: Genetic algorithm based neural networks applied to fault classification for EHV transmission lines with a UPFC, pp. 278–281 (1997)
13. Zhang, B., He, Z., Qian, Q.: Application of wavelet entropy and adaptive nerve-fuzzy inference to fault classification. In: International Conference of Power System Technology, PowerCon, pp. 1–6. IEEE (2006)
14. Youssef, O.A.: An optimised fault classification technique based on Support-Vector-Machines. Paper presented at the Power Systems Conference and Exposition, PSCE 2009. IEEE/PES (2009)
15. Bouloutas, A.T., Calo, S., Finkel, A.: Alarm correlation and fault identification in communication networks. IEEE Transactions on Communications 42(234), 523–533 (1994)
16. Shahid, N., Aleem, S., Naqvi, I.H., Zaffar, N.: Support Vector Machine based fault detection & classification in smart grids. Paper presented at the Globecom Workshops (GC Wkshps). IEEE (2012)
17. Mahamedi, B.: A novel setting-free method for fault classification and faulty phase selection by using a pilot scheme. In: 2nd International Conference of the Electric Power and Energy Conversion Systems (EPECS) (2011)
18. Chutani, S., Decotignie, J.D.: A perspective on fault diagnosis of industrial communication networks. In: Proceedings of the IEEE International Workshop of the Factory Communication Systems, WFCS 1995 (1995)

19. Yang, Q., Guo, J., Zhang, D., Liu, C.: Fault Diagnosis Based on Fuzzy C-means Algorithm of the Optimal Number of Clusters and Probabilistic Neural Network. International Journal of Intelligent Engineering & Systems 4(2), 51–59 (2011)
20. Dashora, K., Kriti, P., Dalal, P., Panwar, D.A.: Software Fault Prediction Using Fuzzy C-Means Clustering and Feed Forward Neural Network. International Journal of Digital Application & Contemporary Research 2(1) (2013)
21. Elbanby, G., El Madbouly, E., Abdalla, A.: Fuzzy principal component analysis for sensor fusion. In: 11th IEEE International Conference in Information Science, Signal Processing and their Applications (ISSPA), pp. 442–447. IEEE (2012)
22. Qader, K., Adda, M.: Network Faults Classification Using FCM. Paper Presented at the 17th International Conference on "Distributed Computer and Communication Networks (DCCN-2013):" Control, Computation, Communication, Moscow (2013)
23. Wu, D., Yang, Q., Tian, F., Zhang, D.X.: Fault Diagnosis Based on K-Means and PNN. In: 3rd International Conference on Intelligent Networks and Intelligent Systems (ICINIS), pp. 173–176. IEEE (2010)
24. Specht, D.F.: Probabilistic neural networks. Neural Networks 3(1), 109–118 (1990)
25. Shahsavarani, S.: Probabilistic Neural Network (2012),
    http://cse-wiki.unl.edu/wiki/index.php/
    Probabilistic_Neural_Network#Architecture

Estimation of the Electric Field across Medium Voltage Surge Arresters Using Artificial Neural Networks

Lambros Ekonomou[1], Christos A. Christodoulou[2], and Valeri Mladenov[3]

[1] City University London, School of Engineering and Mathematical Sciences,
Department of Electrical and Electronic Engineering, London EC1V 0HB, United Kingdom
[2] National Technical University of Athens, School of Electrical and Computer Engineering,
High Voltage Laboratory, 9 Iroon Politechniou St., Zografou Campus, 157 80 Athens, Greece
[3] Department of Theoretical Electrical Engineering,
Technical University of Sofia, Sofia 1000, "Kliment Ohridski" blvd. 8, Bulgaria
lambros.ekonomou.1@city.ac.uk, christ_fth@yahoo.gr,
valerim@tu-sofia.bg

Abstract. Artificial neural networks (ANNs) are addressed in order to estimate the electric field across medium voltage surge arresters, information which is very useful for diagnostic tests and design procedures. Actual input and output data collected from hundreds of measurements carried out in the High Voltage Laboratory of the National Technical University of Athens (NTUA) are used in the training, validation and testing process. The developed ANN method can be used by laboratories and manufacturing/retail companies dealing with medium voltage surge arresters which either face a lack of suitable measuring equipment or want to compare/verify their own measurements.

Keywords: Artificial neural networks, electric field, measurements, surge arresters.

1 Introduction

Nowadays artificial neural networks (ANNs) are being applied to an increasing number of real-world problems of considerable complexity due to their computational speed, their ability to handle complex non-linear functions, robustness and great efficiency, even in cases where full information for the studied problem is absent. Many interesting ANN applications have been reported also in power system areas [1], where they are widely used in short term load forecasting, in fault classification and fault location in transmission lines [2-5], in voltage stability analysis [6], in power system economic dispatch solution problems and in power system stabilizer design [1]. Furthermore the ANNs present to have applications in the solution of the power flow problem [7], to the effective distance protection of the transmission lines [8, 9], to the prediction of high voltage insulators' flashover [10] and to the calculation of insulators' surface contamination under various meteorological conditions [11]. Finally studies, which are using ANNs, have been presented for predicting the magnetic performance of strip-wound magnetic cores [12], for the evaluation of

V. Mladenov et al. (Eds.): EANN 2014, CCIS 459, pp. 227–236, 2014.

lightning overvoltages in distributions lines [13] and for the lightning protection of high voltage transmission lines [14].

In this paper artificial neural networks (ANNs) are addressed in order to estimate the electric field across medium voltage surge arresters, information which is very useful for diagnostic tests and design procedures. Actual electric field measurements carried out in the High Voltage Laboratory of the National Technical University of Athens (NTUA) on a medium voltage metal oxide gapless polymeric (silicon rubber) housing surge arrester, are used in order to train, validate and test the proposed ANNs. Several structures, learning algorithms and transfer functions for an ANN multi-layer perceptron network are tested in order to produce the ANN models with the best generalising ability. The developed ANN method can be used by laboratories, which are facing either a lack of suitable measuring equipment or want to compare the results with their own measurements.

2   Surge Arresters

Surge arresters are semiconductor devices, which are used in electrical power systems in order to protect them against lightning and switching overvoltages. Arresters are installed between phase and earth and act as bypath for the overvoltage impulse, since they are designed to be insulators for nominal operating voltage, conducting at most few milliamperes of current and good conductors, when the voltage of the line exceeds design specifications to pass the energy of the overvoltage wave to the ground. Even though a great number of arresters, which are gapped arresters with resistors, made of SiC are still in use, the arresters installed today are almost all metal oxide arresters without gaps, which means arresters with resistors made of metal oxide [15]. The distinctive feature of a metal oxide arrester is its extremely nonlinear V–I characteristic, rendering unnecessary the disconnection of the resistors from the line through serial spark gaps, as it is found in the arresters with SiC resistors. Additionally, metal oxide arresters are inherently faster-acting than the gapped type, since there is no time delay due to series air gaps extinguishing the current [16].

The basic parts of a metal oxide surge arresters are the cylindrical metal-oxide resistor blocks, the insulating housing and the electrodes (Fig.1). Between the varistor column and the polymeric housing there is a glassfibre structure, that the electric field around a surge arrester is influenced by the geometry of the arrester and the electrical characteristics of the participating materials [17]. Electric field modelling helps the designers to know and consider the important factors affecting the maximum field intensity in the arrester, avoiding too high potential gradients inside and outside the arrester, especially during the transient conditions, a phenomenon which can cause damages to the arrester insulating system that brings it to a premature failure [18]. Hence, the study and the knowledge of the electric field around an arrester can be useful for diagnostic tests [19] and design procedures. Many researches have computed the electric field around a metal oxide arrester using appropriate simulation toolboxes (PC Opera, Cosmol, etc.), examining different cases, such as surface pollution, broken sheds, etc. [17, 18, 20-22].

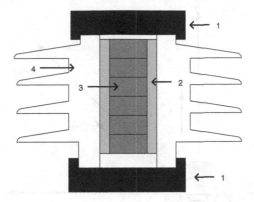

Fig. 1. MO surge arresters cut. (1. Electrodes, 2. Fiberglass, 3. Nonlinear resistor, 4. External insulator).

3   Measurement System

The test arrangement for the measurement of the electric field is shown in Fig. 2. Through a 230V/0…230V variac, a 220V/100kV transformer was fed. The voltage was measured in the primary of the transformer using a digital voltmeter and the applying on the arrester voltages were: 12kV (nominal value), 13.2kV (MCOV) and 16.5kV (rated voltage), which correspond to typical values of the Hellenic distribution network.

In order to measure the electric field around the surge arrester two appropriate calibrated field meters Narda and PMM/8053 were used. The sensors for each instrument, EFA 300 and EHP-50C correspondingly, were placed on an appropriate tripod and were connected via a fiber optic (Fig. 2). Sensors were moved in different directions on the horizontal plane, in various distances along five different axes (Fig. 3) and in various heights. The user of the field meter was at least 10m away from the sensor, in order to avoid interferences to the electric field.

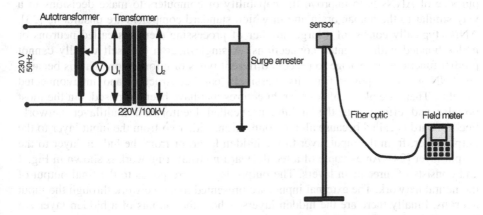

Fig. 2. Experimental set-up used for the measurement of the electric field distribution

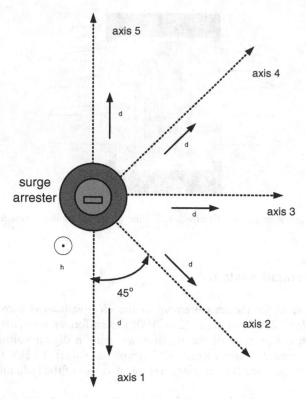

Fig. 3. Experimental set-up used for the measurement of the electric field distribution

4   Artificial Neural Networks (ANN)

Artificial neural networks (ANN) also known as neurocomputers, parallel distributed processors or connectionist models are devices that process information. The main purpose of ANNs is to improve the capability of computers to make decisions in a way similar to the human brain and in which standard computers are unsuitable [23]. ANNs typically consist of a large number of processing elements called neurons or nodes bonded with weighted connections. A single neuron by itself usually cannot predict functions or manage to process different types of information. This is because an ANN gains its power from its massively parallel structure and interconnected weights. There are plenty of ANN architectures that have been proposed, but the most popular and effective is the architecture called feedforward multilayer network. Feedforward is called because all the connections either go from the input layer to the output layer, from the input layer to the hidden layer, or from the hidden layer to the output layer [24]. An example of a feedforward multilayer network is shown in Fig. 4 and consists of three main layers. The output layer corresponds to the final output of the neural network. The external inputs are presented to the network through the input neurons. Finally there are the hidden layers, where the outputs of a hidden layer are the inputs to the following layer.

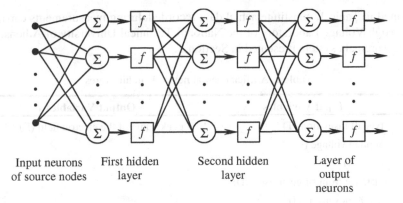

Input neurons   First hidden   Second hidden   Layer of
of source nodes   layer   layer   output
                                                                    neurons

Fig. 4. An example of a feedforward multilayer network

ANNs have the ability of learning by adaptively adjusting their weights using a training algorithm. Ideally, the learning process will adjust the weights of the ANN such that when a given input is presented to the network, the desired output is produced. The learning process starts by presenting the inputs to the network. The set of all the training data is called the training set. The weights are adjusted and the network is expected to produce a particular output for each input. The sum of all weighted inputs represents the neuron transfer function. The set of the desired outputs is called the target set. Each presentation of the training set to the network is called an iteration or epoch.

One of the most popular and powerful learning algorithms used to train feedforward multilayer networks is the backpropagation [25]. The algorithm has two phases, the forward phase and the backward phase. In the forward phase, the input data are presented to the network and the outputs of the hidden layer are given as inputs to the proceeding layer. This process is continued until the final output is computed. In the backward phase, the error signal is calculated and propagated backwards to adjust the weights such that the sum of the squared error is minimized. The mathematical complexity and the time requirements of the algorithm increase by increasing the number of neurons in the hidden layers. Thus, each ANN is determined according to its architecture, the transfer function and the learning algorithm [26].

5   Design, Training, Validation and Testing of the Proposed ANN

The goal of this work is to develop an ANN capable to accurately estimate the electric field across medium voltage surge arresters. A feedforward multilayer network and the backpropagation learning algorithm have been selected for this purpose. Five parameters that play an important role in the estimation of the electric field across medium voltage surge arresters are considered as the inputs to the neural network, while as output the value of the electric field is considered. These data, which are presented in Table 1, constitute actual data recorded during measurements carried out in the High Voltage Laboratory of the National Technical University of Athens, using the measurement system presented in Section 3.

Table 1. Artificial neural network architectures

Input Variables	Output Variables
- type of insulation (T)	- peak value of the electric field (E)
- applied voltage (U)	
- axes (A)	
- distance from surge arrester (D)	
- height of sensor (H)	

More specifically several hundreds of measurements have been performed using the two different field meters. This extensive number of measurements is due to the many parameters. The input parameters were the: three different types of arresters' insulation (T) (polymeric, porcelain, silicon), three different applying voltages (U) (nominal, MCOV and rated), five different axes (A), six distances (D) from the surge arrester (0.5m, 0.8m, 1.1m, 1.4m, 1.7m, 2m) and three different heights of sensors (H) (13cm, 21cm, 29cm). The output parameter of the ANN was the peak value of the electric field (E).

The structure of the developed network, i.e., the number of hidden layers and the number of nodes in each hidden layer, was decided by trying several varied combinations in order to select the structure with the best generalizing ability amongst the all tried combinations, considering that one hidden layer is adequate to distinguish input data that are linearly separable, whereas extra layers can accomplish nonlinear separations [27, 28]. This approach was followed in this work, since the selection of an optimal number of hidden layers and nodes for a feedforward network is still an open issue, although several papers have been published in these areas.

As it is mentioned earlier each ANN is also determined according to the transfer function and the learning algorithm that it uses. In this work three different transfer functions (the hyperbolic tangent sigmoid, the logarithmic sigmoid and the hard limit) and three different training functions of the backpropagation learning algorithm (the Gradient Descent, the Gradient Descent Momentum with an Adaptive Learning Rate and the Levenberg-Marquardt), were examined in order to be selected this one, that contributes to the best ANN's generalizing ability.

The proposed ANN was trained using the MATLAB Neural Network Toolbox [29]. 1620 of each input and output data, were used to train and validate the artificial neural network. These data refer to measurements conducted with both field meters, in every possible combination of: a) applying voltage (3 different voltages), b) axis (5 different axes), c) distance from the surge arrester (6 different distances) and d) height of the sensor (3 different heights). In each training iteration, 20% of random data (324 data sets) were removed from the training set and a validation error was calculated for these data. The training process was repeated until a root mean square error between the actual output (value of electric field) and the desired output reach the goal of 0.5% or a maximum number of epochs (it was set to 10,000), is accomplished. Finally, the estimated values of the electric field were checked with the values obtained from situations encountered in the training, i.e., the 1620 values, and others which have not been encountered (additional sets of input and output data, except the 1620, were used).

In order to be found the best architecture for the network, a feedforward multilayer network has been used, each one of the three different training functions and transfer functions were tried and sets of scenarios were taken with inner change of hidden layers (1, 2 or 3) and number of neurons in each hidden layer (2 to 30).

After extensive simulations with all possible combinations of the 3 training functions, the 3 transfer functions, the 1 to 3 hidden layers and the 2 to 30 neurons in each hidden layer it was found that the ANN with 2 hidden layers, with 16 and 18 neurons in each one of them, with the Levenberg-Marquardt training function and the logarithmic sigmoid transfer function has presented the best generalizing ability, had a compact structure, a fast training process and consumed lower memory than all the other tried combinations. The mean square error was minimized to the final value of 0.005 within 9,249.

6   Results

The trained ANN for the estimation of the electric field across medium voltage surge arresters has been applied to ten different case studies (different insulation type, applying voltage, axes, distances from the surge arrester and heights of sensors), which were not part in the training, validation and testing processes and are shown in Table 2. The produced ANN results, which are presented in Table 2, have been compared to actual values of electric field (which are also presented in Table 2), measured during experiments performed in the NTUA's High Voltage Laboratory for exactly the same parameters.

Table 2. Measured electric field versus artificial neural network's results

No.	Varying parameters					E (V/m)	
	T	U (kV)	A	D (m)	H (cm)	Measured	ANN
1	polym	12	1	0.8	21	1671.7	1736.8
2	polym	13.2	2	1.4	29	528.1	514.5
3	polym	16.5	3	2	29	225.6	277.1
4	porcel	12	2	1.1	13	907.5	927.3
5	porcel	12	3	1.7	21	327.1	298.1
6	porcel	13.2	4	0.5	29	4250.3	4499.4
7	porcel	16.5	5	2	21	218.8	198.3
8	silicon	12	1	1.7	21	335.0	304.8
9	silicon	13.2	3	0.8	13	1675.3	1602.7
10	silicon	16.5	5	1.4	13	543.2	524.6

The results obtained using the proposed ANN are very close to the actual measured ones, something which clearly implies that the proposed ANN method is well working and has an acceptable accuracy, since the measured maximum electric and magnetic field strengths are close enough to these calculated by the designed ANN model. Fig. 5 presents the percentage error between actual measured and ANN's results.

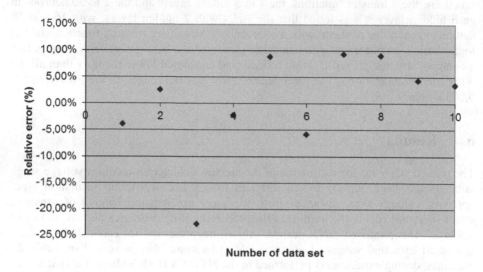

Fig. 5. Relative error for the E between measured and estimated, using the ANN, values

7   Conclusions

The paper describes in detail an artificial neural network for the estimation of the electric field across medium voltage surge arresters. A feed-forward artificial neural network was used and several different learning algorithms, transfer functions and structures were considered in an effort to be selected the ANN model which presented the best generalizing ability, had a compact structure, a fast training process, consumed lower memory and represented the problem accurately among the all tried combinations. The results of the developed ANN model proved its accuracy, since they are very close to the actual measured ones, something which clearly implies that the proposed ANN method is well working. The proposed ANN method, can be used by laboratories and manufacturing/retail companies dealing with medium voltage surge arresters which either face a lack of suitable measuring equipment or want to compare/verify their own measurements.

References

1. Aggarwal, R., Song, Y.: Artificial neural networks in power systems. III Examples of applications in power systems. Power Engin. Journal 12(6), 279–287 (1998)
2. Mahanty, R.N., Gupta, P.B.D.: Application of RBF neural network to fault classification and location in transmission lines. IEE Proc-Gen Tran. Distr. 151(2), 201–212 (2004)
3. Mazon, A.J., Zamora, I., Gracia, J., Sagastabeutia, K.J., Saenz, J.R.: Selecting ANN structures to find transmission faults. IEEE Computer Appl. in Power 14(3), 44–48 (2001)
4. Vasilic, S., Kezunovic, M.: An improved neural network algorithm for classifying the transmission line faults. Power Engin. Society Winter Meeting 2, 918–923 (2001)
5. Gardoso, G., Rolim, J.G., Zurn, H.H.: Application of neural-network modules to electric power system fault section estimation. IEEE Trans. on PWRD 19(3), 1034–1041 (2004)
6. Schmidt, H.P.: Application of artificial neural networks to the dynamic analysis of the voltage stability problem. IEE Proc-Gen Tran. Distr. 144(6), 371–376 (1997)
7. Paucar, V.L., Rider, M.J.: Artificial neural networks for solving the power flow problem in electric power systems. Electric Power Systems Research 62, 139–144 (2002)
8. Dash, P.K., Pradhan, A.K., Panda, G.: Application of minimal radial basis function neural network to distance protection. IEEE Trans. on PWRD. 16(1), 68–74 (2001)
9. Coury, D.V., Jorge, D.C.: Artificial neural network approach to distance protection of transmission lines. IEEE Trans. on PWRD 13(1), 102–108 (1998)
10. Cline, P., Lannes, W., Richards, G.: Use of pollution monitors with a neural network to predict insulator flashover. Electric Power Systems Research 42, 27–33 (1997)
11. Ahmad, A.S., Ghosh, P.S., Aljunid, S.A.K., Said, H.A.I., Hussain, H.: Artificial neural network for contamination severity assessment of high voltage insulators under various meteorological conditions. In: AUPEC, Perth (2001)
12. Miti, G.K., Moses, A.J.: Neural network-based software tool for predicting magnetic performance of strip-wound magnetic cores at medium to high frequency. IEE Proc-Sci. Meas. Technol. 151(3), 181–187 (2004)
13. Martinez, J.A., Gonzalez-Molina, F.: Statistical evaluation of lightning overvoltages on overhead distribution lines using neural networks. Power Engin. Society Winter Meeting 3, 1133–1138 (2001)
14. Sidhu, T.S., Singh, H., Sachdev, M.S.: Design, implementation and testing of an artificial neural network based fault direction discrimination for protecting transmission lines. IEEE Trans. on PWRD 10(2), 697–706 (1995)
15. Hinrichsen, V.: Metal-oxide surge arresters, 1st edn. Siemens (2001)
16. James, R.E., Su, Q.: Condition assessment of high voltage insulation in power system equipment, 1st edn. IET Power and Energy Series, p. 53 (2008)
17. Vahidi, B., Nasab, R.S., Moghani, J., Sh., K.S.A., Hosseinian, S.H.: Three dimensional analyses of electric field and voltage distribution on ZnO surge arrester with broken sheds. In: 2005 IEEE/PES Trans. and Distrib. Conf. & Exhib.: Asia and Pacific, Dalian, China (2005)
18. Meshkatoddini, M.R.: Study of the electric field intensity in bushing integrated ZnO surge arresters by means of finite element analysis. In: COSMOL Users Conf., Boston (2006)
19. Lundquist, J., Stenstrom, L., Schei, A., Hansen, B.: New method of the resistive leakage currents of metal-oxide surge arresters in service. IEEE Trans. on PWRD 5(4), 1811–1822 (1990)
20. Vahidi, B., Nasab, R.S., Moghani, J.S.: Analysis of electric field and voltage distributions on ZnO surge arrester for polluted condition. In: XIV Int. Symp. on High Voltage Engin., Tsinghua University, Beijing, China (2005)

21. Karthik, R.: A novel analysis of voltage distribution in zinc oxide arrester using finite element method. Int. J. of Recent Trends in Engineering 1(4), 1–3 (2009)
22. Han, S.J., Zou, J., Gu, S.Q., He, J.L., Yuan, J.S.: Calculation of the potential distribution of high voltage metal oxide arrester by using an improved semi-analytic finite element method. IEEE Trans. on Magnetics 41(5), 1392–1395 (2005)
23. Abe, S.: Neural networks and fuzzy systems. Kluwer Academic Publishers, Boston (1997)
24. Haykin, S.: Neural Networks: a comprehensive foundation. MacMillan College Publishing Company, New York (1994)
25. Maghami, P.G., Sparks, D.W.: Design of neural networks for fast convergence and accuracy: dynamics and control. IEEE Trans. on Neural Networks 11(1), 113–123 (2000)
26. Nolles, O.: Nonlinear system identification: from classical approaches to neural networks and fuzzy models. Springer, Berlin (2001)
27. Lippmann, R.: An introduction to computing with neural nets. IEEE ASSP Magazine 4(2), 4–22 (1987)
28. Tamura, S.I., Tateishi, M.: Capabilities of a four-layered feedforward neural network: four layers versus three. IEEE Trans. on Neural Nets 8(2), 251–255 (1997)
29. Demuth, H., Beale, M.: Neural network toolbox user's guide for use with MATLAB (2002)
30. Hagan, M.T., Demuth, H.P., Beale, M.: Neural network design. PWS Publishing, Boston (1996)

Decoding Hand Trajectory from Primary Motor Cortex ECoG Using Time Delay Neural Network[*]

Abdessalam Kifouche[1,2,3,**], Vincent Vigneron[2,**],
Mohammad B. Shamsollahi[4], and Abderrezak Guessoum[1]

[1] LATSI, University of Blida, Algeria
[2] IBISC, University of Evry, France
[3] University of Ghardaia, Algeria
[4] BiSIPL, Sharif University of Technology, Tehran, Iran
abdessalam_kifouche@yahoo.fr, vvigne@iup.univ-evry.fr,
abderguessoum@yahoo.com

Abstract. Brain-machines - also termed neural prostheses, could potentially increase substantially the quality of life for people suffering from
motor disorders or even brain palsy. In this paper we investigate the
non-stationary continuous decoding problem associated to the rat's hand
position. To this aim, intracortical data (also named ECoG for *electro-
corticogram*) are processed in successive stages: spike detection, spike
sorting, and intention extraction from the firing rate signal.

The two important questions to answer in our experiment are (*i*) is
it realistic to link time events from the primary motor cortex with some
time-delay mapping tool and are some inputs more suitable for this mapping (*ii*) shall we consider separated channels or a special representation
based on multidimensional statistics. We propose our own answers to
these questions and demonstrate that a nonlinear representation might
be appropriate in a number of situations.

Keywords: BMI, Time Delay Neural Network, nonlinear regression,
spikes.

1 Introduction

Neural prostheses offer the possibility to translate electrical neural activity from
the brain into control signals for guiding paralyzed upper limbs, prosthetic arms,
or computer cursors. Several research groups have already demonstrated that
monkeys as well as human are capable to learn how to drive a robot arm, or more
generally communicate with the outside world, simply by activating neurons
ensemble that participate in natural arm movements.

---

[*] This project was supported in part by funding from the Hubert Curien program of
the Foreign French Minister and from the Taiwan NSC. The neural activity recordings were kindly provided by the Neuroengineering lab. of the National Chiao-Tung
University.

[**] Corresponding authors.

V. Mladenov et al. (Eds.): EANN 2014, CCIS 459, pp. 237–247, 2014.
© Springer International Publishing Switzerland 2014

Multiple electrode arrays allow neurophysiologists to record the spiking activities of an increasing number of neurons. For instance, they have made feasible the recording of a large number of hippocampal cells along with the rat's position in its environment and hence the quantitative analysis of how rat's brain encode spatial information in short term memory and use it for voluntary or non-voluntary action. Investigating these questions requires a collection of statistical tools to analyze how the animal's position is represented in term of firing patterns of place cells. This is the *decoding problem*.

We process spike trains in order to extract a 'firing rate'; the emphasis is on getting things to work robustly, with minimal efforts and with minimal delays, since the decoding must be real time. Estimation algorithms can be designed to decode the desired trajectory from the neural activity patterns. A control system could then generate appropriate signals for continuously guiding a paralyzed or prosthetic arm through space. Such control is indeed a daunting ultimate goal, but would provide a presumably natural control suitable for clinically viable systems.

The core part is the spike train generated from various neurons. Sorting correctly the spikes with respect to their source improve significantly the decoding performance. Hence, before dealing with decoding problem, the spikes should be sorted and the firing rate of each neuron extracted.

In literature, various methods have been introduced to decode brain activities, see *e.g.* [2,13]. This can be formulated systematically in a state space framework: Wu *et al.* in [12] modeled the hand representation (position on the two axes, velocity and acceleration) and the probabilistic relationship between this motion and the firing rates with a Kalman filter (KF). State space methods provide a coherent framework for modeling stochastic dynamical systems. Gage *et al.* in [3] implemented a co-adaptive KF to train a rat for cortical control tasks. This method is able to estimate hidden states despite the lack of an accurate model of the system. Recently Brockwell [1] used particle filters to estimate the hidden states from a sequence of measurements using Monte Carlo algorithm, without any assumption on the distribution of the observations.

We have investigated motor cortex responses recorded during movement in freely moving rats to provide evidence for the relationship between these patterns and special behavioral task, as illustrated Fig. 1(a). The experiment set up at the National Tsin Hua University (NTHU) in Taiwan. With respect to previous works, we focus this time on hand trajectory prediction with a Time Delay Neural Network (TDNN) [4,9].

2 Experimentation and Data Representation

2.1 Animal Training and Behavioral Tasks

The study, approved by the Institutional Animal Care and Use Committee at the National Chiao Tung University, was conducted according to the standards established in the Guide for the Care and Use of Laboratory Animals. Four male Wistar rats weighing 250-300 g (BioLASCO Taiwan Corp., Ltd.) were individually housed on a 12 h light/dark cycle, with access to food and water *ad libitum*.

Dataset was collected from the motor cortex of awake animal performing a simple reward task. In this task, male rats (BioLACO Taiwan Co.,Ltd) were trained to press a lever to initiate a trial in return for a water reward. The animals were water restricted 8-hours/day during training and recording session but food were always provided to the animal *ad lib* every day.

2.2   Chronic Animal Preparation and Neural Ensemble Recording

The animals were anesthetized with pentobarbital (50 mg/kg i.p.) and placed on a standard stereotaxic apparatus (Model 9000, David Kopf, USA). The dura was retracted carefully before the electrode array was implanted. The pairs of 8 microwire electrode arrays (no.15140/13848, 50m in diameter; California Fine Wire Co., USA) are implanted into the layer V of the primary motor cortex (M1). The area related to forelimb movement is located anterior 2-4 mm and lateral 2-4 mm to Bregma. After implantation, the exposed brain should be sealed with dental acrylic and a recovery time of a week is needed.

During the recording sessions, the animal was free to move within the behavior task box (30 cm×30 cm× 60 cm), where rats only pressed the lever via the right forelimb for receiving 1-ml water reward as shown in Fig. 1(a). A Multi-Channel Acquisition Processor (MAP, Plexon Inc., USA) was used to record neural signals. The recorded neural signals were transmitted from the headstage to an amplifier, through a band-pass filter (spike preamp filter: 450-5 kHz; gain: 15,000-20,000), and sampled at 40 kHz per channel as depicted in Fig. 1(b). Simultaneously, the animal's behavior was recorded by the video tracking system (CinePlex, Plexon Inc., USA) and examined to ensure that it was consistent for all trials included in a given analysis [8]. The obtained data were composed of 48 channels (number of neurons) containing succession of '1' separated by long silence of '0'. Another representation is used based on the rate of spike smoothed with a Gaussian window.

3   Temporal Pattern Recognition

3.1   Data Reduction Techniques

Because the rat live in an ever-changing environment, an intelligent system must encode patterns over time, recognizing and generating temporal patterns. A question perhaps as old as modeling is "which variables are important ?". Because the need to select a model applies to more than just variable selection in regression models, there is a rich variety of answers. In the field of brain-machines the selection of relevant features is considered absolutely necessary for the ECoG dataset, since the neural correlates are not known in detail. The time delay between the hand position (output the vector Y_{t+k} to predict) and the intracortical activity (the input vector X_t) is a key point in this work, but cannot be determined exactly or is a shifting value with time.

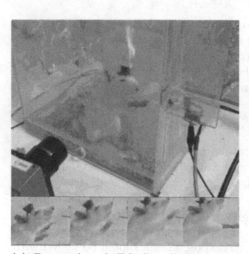

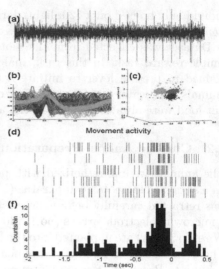

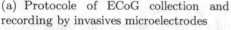

(a) Protocole of ECoG collection and recording by invasives microelectrodes

(b) Spikes sorting and representation

Fig. 1. Experimental setup for neural activities recording and the video captures related animal behavioral task simultaneously

Assume that the training data is given by a set of observations $\{\mathbf{x}(t), \mathbf{y}(t)\}_{t=1}^{N}$, with $\mathbf{x}(t) = [x_1(t), \ldots, x_q(t)]^T \in \mathbb{R}^q$ and $\mathbf{y} \in \mathbb{R}^2$. At time instant t, $\mathbf{u}(t) = [x_1(t), x_1(t - \Delta), \ldots, x_1(t - (n_1 - 1)\Delta), \ldots, x_q(t), \ldots, x_q(t - (n_q - 1)\Delta)]$ is defined as the input set with all variables with all possible time-lags. The goal of the variable and delay selection procedure is to select a subset $S \in \mathcal{X}$ of the most significant variables for the prediction setting. Reducing dimension may also improve the accuracy of an inferred predictive model. Moreover, reducing the number of features may give insights into the working of the system itself. Inference and/or prediction may be computationally costly in high dimensions.

3.2 How Time is Embodied in Temporal Patterns?

By two ways: (i) the *temporal order* which refer to the ordering among the components of a sequence (ii) the *time duration*. How to choose the lag-order for each of the q variables ? Temporal processing is a challenge because the information is embedded in time, not simultaneously available.

Cross-Covariance and Cross-Correlation. The correlation represents how strongly one variable implies the other, based on the available data. Suppose we make N observations on two variables at unit time intervals over the same period and denote a series of observations by $(x_1, y_1), \ldots, (x_n, y_n)$, that may be regarded as a finite realization of a discrete-time bivariate stochastic process

(X_t, Y_t) observed sequentially through time. We use the following notations: the means are resp. $E(X_t) = \mu_X$ and $E(Y_t) = \mu_Y$, and the covariances are resp. $\text{cov}(X_t, X_{t+k}) = \gamma_X(k)$ and $\text{cov}(Y_t, Y_{t+k}) = \gamma_Y(k)$. Then the cross-covariance function is defined by $\text{cov}(X_t, Y_{t+y}) = E[(X_t - \mu_X)(Y_t - \mu_Y)] = \gamma_{XY}(k)$ and is a function of the lag only, because the processes are assumed to be stationary. The cross-correlation function, $\rho_{XY}(k)$, which is defined by

$$\rho_{XY}(k) = \gamma_{XY}(k)/\sqrt{\gamma_X(0)\gamma_Y(0)} = \gamma_{XY}(k)/\sigma_X\sigma_Y, \tag{1}$$

where $\sigma_X = \sqrt{\gamma_X(0)}$ denotes the standard deviation of the X-process, and similarly for σ_Y. This function measures the Bravais-Pearson correlation between X_t and Y_{t+k}. If the cross-correlation ρ_{XY} is 0, then there is no correlation between X and Y meaning that they are independent. For multivariate time series, each variable is regarded as a feature.

A cross-correlation coefficient matrix $M(k)$ is a symmetric matrix, where the (i, j)th entry in the matrix represents the correlation between the ith and jth variables.

Spearsmans Correlation Method. In eq. (1), ρ_{XY} captures only the linear relationship between X and Y. The Spearsmans rank correlation coefficient is a nonlinear measure of statistical dependence between 2 variables, and is computed as the Pearson correlation coefficient between the ranked variables. Let i_X and i_Y denote the value rank for X and Y resp.[1]. The Spearsmans rank correlation coefficient is

$$\rho_S = 1 - \frac{6}{N(N^2 - 1)} \sum_{i=1}^{N} (i_X - i_Y)^2. \tag{2}$$

Entropy Feature-Based Selection Algorithm. As explained before, there are many channels of real recordings available. In prac- tice, it happens that some of the electrodes are detached from the cortex, making there corresponding channel totally noisy. These electrodes should be eliminated before the main processing. The electrode selection strategy originally suggested in [10,11], uses a unique reference signal for the rejection of the channels which contain the most mutual information. The proposed algorithm is based on the mutual information (I) between two signals. The mutual information is defined as the Kullback-Leibler divergence [7] between the joint pdf and the product of the marginal pdf (probability density function):

$$I(X, Y) = \int_{X,Y} p(x, y) \log \frac{p(x, y)}{p(x)p(y)} dx dy. \tag{3}$$

Three major properties of I are: $i)$ it is non-negative, $ii)$ $I(X, Y) = 0$ if and only if X and Y are independent, and $iii)$ $I(X, Y)$ is maximum for $X = Y$.

---

[1] The largest value of X has rank 1, the second largest value rank 2, etc.

The aim of the selection algorithm is to select n signals $(U_1, \ldots, U_n)$ among a set of n signals $(X_1, \ldots, X_n)$. The first signal U_1 must be chosen by another method (*e.g. randomly*). At each step of the algorithm, we choose the signals which are as independent as possible from the already selected signals $U_j, j = 1, \ldots, z-1$, *i.e.* which minimizes the sum of the mutual informations with the U_j's; in other words, X_k is the zth selected signals ($U_z = X_k$) if the following cost function $f_z(i)$ is minimized for $i = k$:

$$f_s(i) = \sum_{j=1}^{z-1} I(X_i, U_j). \tag{4}$$

After the selection of X_k, it is removed from the initial set to avoid an eventual second selection. The selected subset will contain signals which are mutually "quite different", because of the minimisation of the mutual information. From another point of view, selecting signals which minimize the mutual information between them is a good preprocessing, because they have a low dependence level! We stop the algorithm when n' signals are selected ($z = n'$).

To select the best set of features, we have to meet the following constraints: (*i*) a selected feature is irrelevant if it is uncorrelated with the response (*ii*) a selected feature is redundant if it is highly correlated with other features.

3.3   Time-Delay Neural Network

Multilayer perceptrons (MLP) offer a popular approach to temporal pattern learning. MLPs have been demonstrated to be effective for static pattern recognition. It is natural to combine MLP with short term memory model to do temporal pattern recognition. Waibel reported an architecture called Time Delay Neural Networks (TDNN) for spoken phoneme recognition.

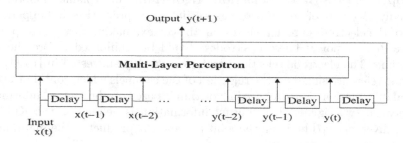

Fig. 2. A time-delay neural network. Only one input $x(t)$ is shown. The delayed inputs $x_i(t - \Delta), \ldots, x_i(t - n_i\Delta), i = 1, \ldots, q$ are fully connected to a hidden layer.

Besides the input layer, TDNN uses two hidden layers and an output layer where each unit encodes one phoneme. The feed-forward connections converge from the input layer to each successive layer so that each unit in a specific layer receives inputs within a limited time window from the previous layer.

Given the raw signal $\mathbf{x}(t)$ the usual delay line technique would be to use $\mathbf{x}(t), \mathbf{x}(t-\Delta), \ldots, \mathbf{x}(t-(n-1)\Delta)$ for the network inputs at time t as sketched in Fig. 2. TDNN will replace the linear model used by Wu *et al.* [12]. The TDNN architecture consists of the delayed versions of the firing counts, which effectively implements a short-term memory mechanism.

However, additional delays in the already high dimensional neuronal input (40 neurons) bring in a huge number of free parameters and training of such large TDNN by simple back-propagation algorithm is inappropriate since the parameter convergence need long iterations and a learning rate to adjust. Hence we propose to use the conjugate gradient algorithm (CGA) to accelerate the weight update.

Algorithm 1. Standard CG algorithm

Require: initialize $\mathbf{w}^{(0)}$ (can be 0);
Ensure: $\mathbf{r}^{(0)} = \mathbf{p}_0 \leftarrow -\nabla C(\mathbf{w}^{(0)})$;
1: $k \leftarrow 0$
2: **repeat**
3:     $\alpha^{(k)} \leftarrow \frac{\mathbf{r}^{(k)T}\mathbf{r}^{(k)}}{\mathbf{p}^{(k)T}A\mathbf{p}^{(k)}}$
4:     $\mathbf{w}^{(k+1)} \leftarrow \mathbf{w}^{(k)} + \alpha^{(k)}\mathbf{p}^{(k)}$
5:     $\mathbf{r}^{(k+1)} \leftarrow \mathbf{r}^{(k)} - \alpha^{(k)}A\mathbf{p}^{(k)}$
6:     $\beta^{(k)} \leftarrow \frac{\mathbf{r}^{(k+1)T}\mathbf{r}^{(k+1)}}{\mathbf{r}^{(k)T}\mathbf{r}^{(k)}}$
7:     $\mathbf{p}^{(k+1)} \leftarrow \mathbf{r}^{(k+1)} + \beta^{(k)}\mathbf{p}^{(k)}$
8:     $k \leftarrow k + 1$
9: **until** $\mathbf{r}^{(k+1)}$ is sufficiently small.
10: **return** $\mathbf{w}^{(k+1)}$

Algorithm 2. Preconditioned CG

Require: initialize $\mathbf{w}^{(0)}$ (can be 0);
Ensure: $\mathbf{r}^{(0)} \leftarrow -\nabla C(\mathbf{w}^{(0)})$;
Ensure: $\mathbf{z}_0 \leftarrow M^{-1}\mathbf{r}_0$; $\mathbf{p}_0 \leftarrow \mathbf{z}_0$;
1: $k \leftarrow 0$
2: **repeat**
3:     $\alpha^{(k)} \leftarrow \frac{\mathbf{z}^{(k)T}\mathbf{r}^{(k)}}{\mathbf{p}^{(k)T}A\mathbf{p}^{(k)}}$
4:     $\mathbf{w}^{(k+1)} \leftarrow \mathbf{w}^{(k)} + \alpha^{(k)}\mathbf{p}^{(k)}$
5:     $\mathbf{r}^{(k+1)} \leftarrow \mathbf{r}^{(k)} - \alpha^{(k)}A\mathbf{p}^{(k)}$
6:     $\mathbf{z}^{(k+1)} \leftarrow M^{-1}\mathbf{r}^{(k+1)}$
7:     $\beta^{(k)} \leftarrow \frac{\mathbf{z}^{(k+1)T}\mathbf{r}^{(k+1)}}{\mathbf{z}^{(k)T}\mathbf{r}^{(k)}}$
8:     $\mathbf{p}^{(k+1)} \leftarrow \mathbf{z}^{(k+1)} + \beta^{(k)}\mathbf{p}^{(k)}$
9:     $k \leftarrow k + 1$
10: **until** $\mathbf{r}^{(k+1)}$ is sufficiently small.
11: **return** $\mathbf{w}^{(k+1)}$

---

4   Learning Algorithm

An error back-propagation learning rule coupled with preconditioned conjugate gradient is used to trained the TDNN weights and minimize the Mean Square Error (MSE) criterion:

$$C(\mathbf{w}) = E[e^2(t)], \tag{5}$$

where $e(t)$ is the difference between the expected signal $y^*(t)$ and model output $y(t) = f(\mathbf{w}, \mathbf{x}(t))$ at time t. The weight vector update is given by

$$\mathbf{w}^{(k+1)} = \mathbf{w}^{(k)} - \eta^{(k)}\nabla C(\mathbf{w}^{(k)}), \tag{6}$$

where $-\nabla C(\mathbf{w}^{(k)})$ is the gradient vector and $\eta^{(k)}$ is the learning rate that should be chosen carefully to make the algorithm stable and converge. But since the gradient algorithm is very sensitive to η, a conjugate gradient algorithm is generally a better choice and results in a faster convergence to reach the minimum of the cost function (5). CG algorithm [5] choose the search direction from the second order approximation $C(\mathbf{w}+\mathbf{h}) \approx C(\mathbf{w})+\nabla C(\mathbf{w})^T\mathbf{h}+\frac{1}{2}\mathbf{h}^T C(\mathbf{w})\mathbf{h}$, where $\mathbf{h}$ is the line search direction. The CG algorithm updates the weights along the conjugate gradient direction. From the current search direction, the next search direction is determined so that it is conjugate to previous search directions. When the CG methods are applied to non quadratic functions, the formula given below, called the Hestenes-Stiefel formula, is considered superior for $\beta^{(k)}$.

In most cases, preconditioning is necessary to ensure fast convergence of the conjugate gradient method that takes the following form in Algo. 2. The preconditioner matrix M has to be symmetric positive-definite and fixed.

5   Methods

5.1   Experiment and Performance Assessment

After the spike sorting stage, we obtained a 48 channels of ECoG represented as a long suit of pulses '1' separated by a long silence '0'. Many techniques of data representation were proposed to carry out the hidden information situated in the inter-spike distances. Most of these techniques are summarized in [6].

In our work we used a moving window of 100 ms time width to compute the spike rate and also of the used camera frequency to follow the lever and get simultaneously position with neural activity rate. Correlation coefficients are calculated and used to select which channels can be used as feature of the machine learning and to eliminate the redundant channels and non correlated channels to the output. Maximal cross-correlation values indicate give the delay of the explanatory variables that will be used as network inputs.

The TDNN model is shown in Fig. 2. The input layer includes in this experiment 10 delayed cortical signals from the 3 most prominent neurons (among 46 neurons having been recorded), between 2-7 hidden neurons with hyperbolic tangent ceil function.

We trained the TDNN by the 2 algorithms as described in section 4. The 2 output values (the 2 dimensional hand position) are centered and reduced to follow a Gaussian law $\mathcal{N}(0,1)$. The layer's weights are initialized randomly between $[-1, +1]$. In both algorithms, we set the learning rate is fixed to $\eta = 0.05$. The training was terminated when the cross validation error continuously increased for more than 10 steps. The network specifications are listed in Tab. 1. While the choice for the right architecture is mainly intuitive and implies arbitrary decisions, an attempt to apply ANN directly fails due to the dimensionality of the inputs. Therefore the dimension of the inputs has been reduced drastically by feature selection.

Tab. 1 illustrates the effects of variation of the number of units in the TDNN on decoding performance quantified for each model by the mean AIC across the 50 data sets for the two decoding algorithms.. Only 10% of the data (7 peaks)

Table 1. Average AIC

Architecture	# param.	CGA		preconditioned CGA	
		x	y	x	y
27-2-2	62	3.691	4.723	2.700	3.720
27-3-2	92	3.394	**0.426**	**1.406**	3.435
27-4-2	122	**2.732**	2.822	2.762	**2.799**
27-5-2	152	5.066	5.148	3.089	5.132
27-6-2	182	5.105	6.162	8.117	6.133
27-7-2	212	6.531	9.570	11.541	9.550

are shown here for clarity. The model output (denoted by red lines) can track the x and y coordinate of hand position trajectory most of the time and in particular can catch the peaks well (in blue) if the complex is sufficiently complex. Figures 3.b-g zoom onto the fit in the seven peak region.

The measure of performance is the Akaike information criterion (AIC). As such, AIC provides a means for model selection. The AIC value is $AIC = 2p - 2\ln(L)$, where p is the number of parameters in the model, and L is the maximized value of the likelihood function for the model. AIC deals with the trade-off between the goodness of fit of the model and the complexity of the model. The preferred model is the one with the minimum AIC value.

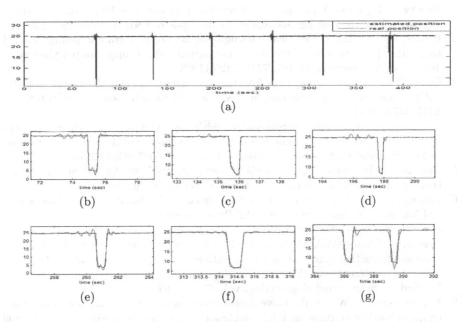

Fig. 3. Hand position tracking with TDNN. Model output (in red) and desired output (in blue).

The number of units required for successful decoding has an important impact on the overall structure of the BMI since it determines the number of electrodes, and channels of the signal conditioning circuitry.

The preconditioned CGA was superior to the classical CGA when reasonable numbers of units (i.e., more than 3 units) were used for the decoding. Here in Fig. 3, the result obtained from the MLP 27-5-2 is shown, i.e., we used the trial-and-error method for the training of the MLP. The performance of the MLP 27-5-2 was much worse than that of the CGA. The performance differences between the CGA andpreconditioned CGA were larger above 5 units.

6   Conclusion and Future Works

A TDNN architecture was proposed for fast hand trajectory decoding from primary motor cortex ECoG. Since the application of TDNN in BMIs yields too many free parameters, the approach performs faster with the conjugate agradient algorithm. However, a TDNN with 20 hidden neurons and 10 time delays of 46 channels of neuronal signal produces more than 1,160 free parameters and we only have 10,000 samples for training and cross-validation. Another issue is the possibility to use a Kalman filter for detecting pulses and grouping them, hence simplifying the activation measurement.

References

1. Brockwell, E., Rojas, L., Kass, R.E.: Recursive bayesian decoding of motor cortical signals by particle filtering. Journal of Neurophysiology 91(4), 1899–1907 (2004)
2. Brown, E.N., et al.: A statistical paradigm for neural spike train decoding applied to position prediction from ensemble firing patterns of rat hippocampal place cells. Journal of Neuroscience 18(18), 7411–7425 (1998)
3. Gage, G.J.: Co-adaptive kalman filtering in a naïve rat cortical control task. In: IEEE Conference on Engineering in Medicine and Biology Conference, vol. 6, pp. 4367–4370 (2004)
4. Ghanbari, A., et al.: Neural spike sorting with a self-training semi-supervised support vector machine. In: Annual International Conference of the IEEE Engineering in Medicine and Biology Society, Osaka, Japan, pp. 6007–6010 (2013)
5. Hestenes, M.R., Stiefel, E.: Methods of conjugate gradients for solving linear systems. Journal of Research of the National Bureau of Standard 49(6) (1952)
6. Dayen, P., Abbott, L.F.: Theoretical Neuroscience. Computational and Mathematical Modeling of Neural Systems. MIT Press (2001)
7. Sameni, R., Vrins, F., Parmentier, F., Herail, F., Vigneron, V., Verleysen, M., Jutten, C., Shamsollahi, M.B.: Electrode selection for noninvasive fetal electrocardiogram extraction using mutual information criteria. In: MaxEnt2006 Proceedings - 26th International Workshop on Bayesian Inference and Maximum Entropy Methods in Science and Engineering, vol. 872, pp. 97–104 (July 2006)
8. Van Staveren, G.W., et al.: Wave shape classification of spontaneaous neural activity in cortical cultures on micro-electrode arrays. In: Proceedings 24th Annual Conference of the EMBS/BMES Society, TX, USA, October 23-26, vol. 3, pp. 2010–2011 (2002)

9. Vigneron, V., Chen, H., Chen, Y.-T., Lai, H.-Y., Chen, Y.-Y.: Decomposition of EEG signals for multichannel neural activity analysis in animal experiments. In: Vigneron, V., Zarzoso, V., Moreau, E., Gribonval, R., Vincent, E. (eds.) LVA/ICA 2010. LNCS, vol. 6365, pp. 474–481. Springer, Heidelberg (2010)

10. Vrins, F., Lee, J.A., Verleysen, M., Vigneron, V., Jutten, C.: Improving independent component analysis performances by variable selection. In: NNSP 2003 proceedings - Neural Networks for Signal Processing, Toulouse, France, pp. 359–368 (September 2003)

11. Vrins, F., Vigneron, V., Jutten, C., Verleysen, M.: Abdominal electrodes analysis by statistical processing for fetal electrocardiogram extraction. In: Proceedings of the 2nd International Conference Biomedical Engineering, Innsbruc, Austria, pp. 244–250 (2004)

12. Wu, W.: Inferring hand motion from multi-cell recordings in motor cortex using a kalman filter. In: Workshop on Motor Control in Humans and Robots: On the Interplay of Real Brains and Artifical Devices, pp. 1–8 (2002)

13. Wu, W., et al.: Closed-loop neural control of cursor motion using a kalman filter. In: 26th Annual International Conference of the IEEE Engineering in Medicine and Biology Society, vol. 6, pp. 4126–4129 (2004)

Author Index